Collective Intelligence in
Computer-Based Collaboration

COMPUTERS, COGNITION, AND WORK

A series edited by:
Gary M. Olson, Judith S. Olson, and Bill Curtis

Fox • The Human Tutorial Dialogue Project: Issues in the Design
of Instructional Systems

Moran/Carroll (Eds.) • Design Rationale: Concepts, Techniques, and Use

Oppermann (Ed.) • Adaptive User Support: Ergonomic Design of Manually and
Automatically Adaptable Software

Smith • Collective Intelligence in Computer-Based Collaboration

Collective Intelligence in Computer-Based Collaboration

John B. Smith
The University of North Carolina

 LAWRENCE ERLBAUM ASSOCIATES, PUBLISHERS
1994   Hillsdale, New Jersey   Hove, UK

Lawrence Erlbaum Associates, Inc., Publishers
365 Broadway
Hillsdale, New Jersey 07642

Cover design by Kate Dusza

Library of Congress Cataloging-in-Publication Data

Smith, John B., 1940–
    Collective intelligence in computer-based collaboration / John
B. Smith
        p.   cm.
    Includes bibliographical references and index.
    ISBN 0-8058-1319-5.—ISBN 0-8058-1320-9 (pbk.)
    1. Work groups—Data processing.   2. Human–computer in-
teraction—Psychological aspects.   3. User interfaces (Computer
systems)—Psychological aspects.   4. Cognitive psychology.   I.
Title.
HD66.S63   1994
658.4'036'0285—dc20   94-13958
                                                               CIP

Books published by Lawrence Erlbaum Associates are printed on
acid-free paper, and their bindings are chosen for strength and dura-
bility.

For Catherine, Ian,
and my other collaborators

Contents

Preface

Like many books, this one has grown in part out of personal experience. For the first 10 or 12 years of my professional life, I tended to work alone. My research interests centered on computer-assisted text and natural language analysis. I was fortunate enough to hold an academic appointment at a large university. At the time, my work was regarded as unusual if not eccentric. It was tolerated by my colleagues, even supported, but none of them shared those interests. Of course, I had professional colleagues at other institutions whom I saw at conferences, and we shared a number of good times talking and arguing ideas. But, for the most part, I had to generate my own motivation and to rely largely on my own ideas.

When I came to the University of North Carolina (UNC) some 10 years ago, one of the joys of that move was the chance to work closely with others. My first collaborators were Steve Weiss, a computer scientist; Marcy Lansman, a cognitive psychologist; Jay Bolter, a classicist; plus five or six graduate students, most of whom were from computer science but also including several from cognitive psychology. Since that time, our group has grown to include some 10 faculty and nearly twice that many graduate students, and we have expanded the multidisciplinary character of the group to include anthropologists.

Our earliest collaborations were concerned with hypermedia computer systems and their application to technical and scientific writing. From the beginning, we believed that if we could understand more clearly the cognitive process of writing, then we should be able to build computer systems consistent with that process. That is, if we could identify key mental activities that comprise expository writing, then we should be able to build corresponding features into our computer systems to support and, we hoped, enhance those same activities.

Our interests in writing and in individual computer users continue, but the focus of our research began to shift several years ago toward issues of collaboration. Thus, what was first a way of working became a topic of research. Of course, our studies of collaboration have gone beyond our own group, but I have benefited greatly by being able to observe as well as participate in an evolving group and to test ideas about collaborative theory against collaborative practice, firsthand.

We were initially drawn in this direction by a desire to better understand collaborative writing and to adapt our writing system so that it could support groups jointly authoring documents. We have since extended the task domain to software development. However, the shift from writing to software development and from individual users to collaborative groups is not so large a step as it may first appear.

Writing documents is a metatask for many different forms of intellectual work. This is true in two respects. First, many intellectual activities ultimately produce a document that records and communicates the results of that activity. For example, when one plans a research project, one normally expresses that plan as a document. Similarly, software development usually involves writing a number of different documents; these include requirements, specifications, the design or architecture of the system, test plans, and user instructions. In fact, it's hard to imagine a substantial intellectual task that doesn't involve writing some form of document as an integral part of that task.

A second reason writing can be considered a metatask lies in the underlying mental processes it draws on and the constraints those processes are subject to. Working out the abstract structure of ideas that is the content of a document is as much a part of the "writing process" as expressing those ideas in words and sentences. Many intellectual tasks involve working out some form of preliminary plan and then building or expressing in detail the individual components that make up the plan. But, in doing so, we inevitably discover problems or subtleties not foreseen during the planning stage. Consequently, we must go back and revise the plan, which, in turn, may lead to additional changes needed to make the rest of the plan consistent with those changes. Thus, the processes of planning, representing, evaluating, and revising are used both iteratively and recursively, as work descends from abstraction into detailed expression or physical realization. As a result, this *overall* process of knowledge-construction is the same, regardless of whether the content is ultimately expressed in words, diagrams, code, or physical material.

Many of these same processes also operate within groups. For example, groups brainstorm together, particularly at the beginning of a new project. They also build plans together, review, and, at times, edit and revise together. But, of course, the individual members that make up the group also work alone between joint work sessions. What is the overall pattern of behavior in a group as its members move back and forth from individual to collective work and from one process to another? If we expect groups to develop products that are coherent, internally consistent, and have intellectual integrity, then we need to understand this overall collaborative process.

What we would like to have, eventually, is a process model of collaboration similar to the process models that have been developed for tasks performed by individuals. We would like that model to be sufficiently general so that it applies to different task domains and to groups working in different organizational contexts. If I am correct in believing that conceptual construction tasks of all kinds draw on a common set of underlying processes and constraints, a general theory of collaboration is possible, although it is likely to differ substantially from models of individual cognition.

I do not get so far as actually defining a model of collaboration in this discussion, but that is the direction in which I am headed. What I do hope to accomplish is to sketch in some detail an image of collaboration as an information processing activity and to describe a framework for research that can help us build on one another's work in a way that may eventually lead to such a theory.

The term I use for this view of collaboration is *collective intelligence* (CI). Because much of the discussion is an attempt to define this concept, I do not describe it in more detail here but, instead, comment on several of its attributes. I associate CI with a set of goals or boundary conditions for collaboration. In trying to identify the essential processes and constraints that operate within groups, I am inevitably drawn toward describing patterns of behavior that can help groups produce products that have greater coherence and consistency. Thus, the discussion includes at various points a sense that some particular way of working may be preferable to another. However, I do not mean to advocate any specific behavior. Biomechanics provides an analogy for the fine line I am trying to walk. By analyzing the different muscle, bone, and ligament structures found in the human body, experts have been able to devise training programs that enhance or strengthen these different systems with the result that athletes who follow these regimens can run faster, jump higher, or throw further than they could otherwise. Similarly, if we understood the intellectual and social systems that operate within collaborative groups, we might eventually be able to train groups and build tools to help them so that they could accomplish their tasks better than they could otherwise.   But, because we do not yet have that knowledge, optimization and advocacy must remain topics for future consideration.

In the past, much more emphasis has been placed on the social dimensions of groups than on the cognitive.   For example, prior studies have commonly examined patterns of influence within groups, qualities of leadership, the role of status, and so on.   In this discussion, I have emphasized the cognitive and conceptual aspects of collaboration and the technology to support them.   Thus, the discussion draws heavily on research from cognitive science and computer science.   I have done this because collaborative groups are also fundamentally intellectual when their goal is to produce a conceptual artifact.   Although I have also tried to show that cognitive processes are often closely, even fundamentally entwined with social processes, the discussion is imbalanced, in part to make up for this deficit on the cognitive side.   I hope that future treatments of collaboration will establish the proper balance between the two.

To do so will require integrating theories and research from a number of disciplines, including anthropology, ethnography, management science, organizational theory, economics, sociology, social psychology, speech communication, and the study of small groups.   Although I wish I had had the time and background to synthesize this material, apart from being impractical, I'm not sure, on reflection, it would have been desirable.   Scholars who have spent a lifetime working in these fields can bring a depth of knowledge and discipline-based understanding that no single individual could ever assimilate.   I hope that this discussion will pique their interest, that they will see ways in which they can expand or correct it, and that they will join an on-going *collaborative* enterprise to develop a valid theory of collective intelligence.

I have written this book with two primary audiences in mind. The first is anyone interested in collaboration who would like to think more about how groups operate and how technology may now and in the future affect the thinking, the social interactions, and the products produced by collaborative groups.   The rapid pace of development in computer networks, distributed systems, and communications makes it increasingly possible for people to interact with one another, even when they are widely separated geographically.   When we take into account the increasing tendency of organizations to encourage people to work in more flexible combinations with one another, we can expect collaboration to become the predominant form of intellectual work. Consequently, this first group of potential readers includes just about anyone who does or is interested in intellectual work.

The second group of readers are my colleagues engaged in research and development in collaboration theory, studies, and systems. I have tried to sketch a comprehensive view of collaboration as a kind of intelligent organism. At present, the field of computer-supported cooperative work (CSCW) is driven largely by the technology and by enthusiasm, with the result that our activities and our systems are scattered and piecemeal. I believe it is not too early to attempt a more inclusive view of collaboration, comparable in scale and hope to the shift now taking place in cognitive science as they move from partial models to comprehensive architectures. Of course, we don't have a comparable body of prior research to build on, but we can generate this base of knowledge more quickly and more efficiently if we are aware of how our individual projects fit within a larger whole. What I have offered here is a candidate whole. Thus, my goal is to provide us with some*thing* to talk about, to debate, to correct, and perhaps, in part, to confirm.

There is a third group of readers who may find this discussion of interest. At present, CSCW is a highly multidisciplinary field, drawing concepts and techniques from a number of different disciplines, including computer science, human–computer interaction, cognitive psychology, anthropology, ethnography, sociology, organizational theory, small group theory, composition theory, economics, and, no doubt, other areas of which I am unaware. As a result, scholars and researchers working in these more mature "ancestor" disciplines may find some of the results emerging from CSCW, including this discussion, at least tangentially related to their own interests.

Although the discussion refers to a number of collaboration studies and support systems, it is not intended to be a comprehensive overview of work in the field. Rather, it is a "think piece" that tries to provide a concrete image of an abstract concept. As I gauge the rate of progress in CSCW, I estimate that it could have a useful half-life of about 5 years. When that time comes, I hope someone else will write a similar book that sketches another possible future from that vantage point.

Finally, I want to acknowledge some of those who have contributed to the ideas presented here. I have been fortunate over the past 10 years to have many outstanding collaborators. At this point, it is impossible to disentangle which ideas originated with me and which with them. A more accurate way to think about what took place is that we all participated in a process of intellectual evolution in which a large, amorphous conceptual structure gradually developed. What I have discussed here draws on that structure and is formed from my understanding of that larger community property we all share and own.

My earliest collaborators were Steve Weiss, Marcy Lansman, Jay Bolter, and Gordon Ferguson. I am indebted to them for their contributions to the earlier research on writing and writing systems, described later, out of which grew our later work on collaboration.

That original group was joined by Don Smith, Kevin Jeffay, Dotty Holland, Dana Smith, Peter Calingaert, and Hussein Abdel-Wahab to form the UNC Collaboratory Project. Reports of their work can be found throughout the book.

More recently, the UNC group has expanded to include David Stotts and Prasun Dewan. Although we have worked together only a short time, I want to acknowledge their contributions, along with that of my other UNC colleagues, to the discussion about future collaboration systems and the research that will be needed to achieve them.

During this period, I have been fortunate to work with a number of graduate students. It is they who have written the systems the Project takes credit for. Many of them have also played key roles in both the architecture and the conceptual basis of those systems.

I also want to acknowledge a debt to my colleague, Fred Brooks, from whom I learned the value of thinking hard in order to think simply.

A number of readers made valuable suggestions for improving the manuscript. Thanks to the students who read and discussed an early draft during a seminar on collaboration and to Don Smith for his helpful comments on several sections. Thanks, also, to Jessika Toral who began the index and to Claire Gingell who completed it and helped with many other tasks involved with preparing the manuscript. Brian Ladd prepared many of the illustrations, showing a real flair for the visual, for which I am grateful. I also want to thank Amy Pierce for her willingness to take a chance with this book, for her patience in allowing me to complete it at my own pace, and for her encouragement throughout.

I am particularly indebted to Jan Walker for performing one of her patented very close, very thoughtful readings of the entire manuscript. Her comments led to a number of corrections and improvements. The problems that remain are, of course, the responsibility of the author.

Portions of the research done at UNC and reported in this book were sponsored by the National Science Foundation and the International Business Machines Corporation. I especially want to acknowledge Larry Rosenberg's efforts and leadership in establishing the Coordination Theory and Collaboration Technology program at NSF; without its support, this book and much of the work on which it is based could not have been done. I am also grateful to University of North Carolina for a research leave during which the book was begun.

Finally, I wish to acknowledge my oldest and best collaborator, Catherine Smith. She has lived these ideas with me, contributed to their development, and has never refused to read a section and give me her thoughtful views, even under the most trying circumstances.

John B. Smith

Copyright   Acknowledgments

The author wishes to thank the following publishers for permission to use previously published figures and textual passages:

Figure 4.1 is reprinted by permission of the publishers from Newell, A., & Simon H. A. (1972). *Human Problem Solving.* Englewood Cliffs, NJ: Prentice-Hall, p. 20.

Figure 4.2 is reprinted by permission of the publishers from Newell, A. (1990). *Unified Theories of Cognition.* Cambridge, MA: Harvard University Press, p. 195. Copyright © 1990 by the President and Fellows of Harvard College.

Figure 4.3 is reprinted by permission of the publishers from Anderson, J. R. (1983). *The Architecture of Cognition.* Cambridge, MA: Harvard University Press, p. 19. Copyright © 1983 by the President and Fellows of Harvard College.

Figures 4.4, 4.5, 4.6, and 4.7 are reprinted by permission of the publishers from Hayes, J. R. & Flower, L. S. (1980). Identifying the organization of writing processes. In L. W. Gregg & E. R. Steinberg (Eds.), *Cognitive processes in writing* (pp. 3–30). Hillsdale, NJ: Lawrence Erlbaum Associates, pp. 11, 13, 19–20, 26.

Figure 4.8 is reprinted by permission of the publishers from Card, S. K., Moran, T. P., & Newell, A. (1983). *The psychology of human-computer interaction.* Hillsdale, NJ: Lawrence Erlbaum Associates, p. 26.

Figures 7.6, 7.7, and portions of the description of meeting 2 are reprinted by permission of the author from Blakely, K. D. (1990). *The application of modes of activity to group meetings: A case study* (Tech. Rep. No. TR90–045). Chapel Hill, NC: UNC Department of Computer Science, pp. 19, 28–31.

Chapter 1

Introduction

The notion of *collective intelligence* (CI) is that a group of human beings can carry out a task as if the group, itself, were a coherent, intelligent organism working with one mind, rather than a collection of independent agents.

The idea — referred to by several different terms — has been around for some time, but with recent interest in collaborative and cooperative work, it is being heard more often. Usually, it carries with it a bit of blue sky or is part of a throw-away line. For example, a grant proposal might suggest that the computer system the project is building to support collaborative work might eventually lead to a form of collective cognition by its users. But what, exactly, does that mean? What mode of thinking would constitute collective intelligence? What would be its characteristics? Would we recognize it if we saw it or experienced it?

In this discussion, I examine the idea of collective intelligence in order to try to pin it down and put some flesh on its bones. Thus, I hope to move discussion from a vague *notion* of collective intelligence to a *concept* that is reasonably well-defined. In the long-term, perhaps those working in this field can eventually build a *theory* of collective intelligence that is sufficiently precise so that it can be tested and refined. If such a theory existed, it could have a number of useful consequences. For example, if we really understood how groups of individuals can occasionally and under particular circumstances meld their thinking into a coherent whole, we would have a better idea of how to build computer and communications systems to support them, how to train other groups to work this way, and how to organize projects and institutions to promote this mode of work. I hope this discussion is a first step toward these goals.

Not everyone believes such a theory is possible. For example, Allen Newell (1990) argued that it is impossible for any group to function as a coherent rational agent. His objection, which I discuss in more detail later, is based on the rate at which information can be transferred from one human being to another. He argued that the bandwidth is insufficient to permit the various members of a group to all share the same knowledge — a condition he believed would be required to achieve what I call collective intelligence. Newell's objection is an important one that must be answered. Although I cannot refute his premise, I try to build a path around that roadblock.

Constraints

Group intellectual activities take place on many different scales. I frame the issue narrowly in order to make the discussion as concrete as possible. Perhaps later these constraints can be relaxed and the concept extended to a broader range of groups.

The discussion is limited to *intellectual* groups that are building some type of concrete conceptual object, such as a technical report, a marketing plan, a computer system, a legislative bill, or an airplane design. Excluded from the discussion, then, are groups that are primarily social, those carrying out manual tasks, or collaborations that produce aesthetic objects.

The discussion is limited to groups that range in size from three to four individuals to a handful of such groups working together on a single project. Excluded, then, are two-person collaborations that often have highly personal dynamics that don't extend to larger groups and, at the other extreme, projects that involve hundreds or even thousands of people. However, within this band of 3–30 people we can consider many of the problems encountered by groups of all sizes as well as the first extrapolation from a single group to a collection of groups.

Third, the discussion is limited to groups working on tasks that last from several weeks to several years. Durations within this range are long enough to raise problems of conceptual coordination and integration of ideas and materials, yet they are sufficiently bounded that the work is not viewed as ongoing and the group becomes institutionalized or bureaucratic in its operations.

Fourth, I distinguish between *collaboration* and *cooperation*. Collaboration carries with it the expectation of a singular purpose and a seamless integration of the parts, as if the conceptual object were produced by a single good mind. For example, a well-done collaborative document has a clear purpose or message. The reader is unable to tell from internal cues which chapters or sections were written by which authors. The sections are also consistent with one another, and one section shows appropriate awareness of the contents of the other sections.

Cooperative work is less stringent in its demands for intellectual integration. It requires that the individuals that comprise a group or, for larger projects, a set of groups carry out their individual tasks in accord with some larger plan. However, in a cooperative structure, the different individuals or groups are not required to know what goes on in the other parts of the project, so long as they carry out their own assigned tasks satisfactorily.

For example, the various teams of biologists that are currently mapping the human genome normally concentrate their research on a single chromosome or portion of a chromosome (DHHS, 1990). Although it could be advantageous, one team does not necessarily have to monitor work going on in other portions of the DNA structure in order to achieve its goals, nor is one team required to reconcile its methods and results with those of other groups. Such integration may eventually come — indeed, we see glimpses of this as newly articulated genes are mapped against various diseases and abnormalities. But, for now, although work within groups may be collaborative, work among groups in this field is more separate and diverse, albeit still cooperative.

I make this distinction between cooperation and collaboration to further limit the discussion. It seems to me far easier to imagine a concept of collective intelligence existing within a collaborative project than in one that is cooperative. Indeed, I suggest that collective intelligence is a requirement for effective collaboration, at least as a goal or boundary condition. Consequently, I limit the rest of this discussion to collaborative groups.

To summarize the constraints outlined so far, I examine a concept of collective intelligence by considering how collaborative groups ranging in size from 3 to 30 individuals working together for periods of several weeks to several years can produce an intellectual product that represents accomplishment of the group's main goal so that the product has the characteristics we would expect had it been produced by a single good mind.

I make one final assumption. The discussion is constrained to collaborative groups that use a computer and communication system as an integral part of their work.   A requirement for collective intelligence is achieving a critical level of coherence in the work of the group. Although I admit the possibility in the abstract that a group might achieve this level of coherence without using a computer system, I cannot personally envision how large groups could coordinate their efforts and integrate the products develop by their individual members to the degree required for CI without such a system. Thus, I assume that CI is a form of intellectual behavior that is at least partially enabled by the technology. Later, when we understand the phenomenon better, it may be possible to relax this constraint and observe or develop CI in groups working without computer assistance.

Intelligence   Amplification

The view of collective intelligence as a form of behavior made possible by some form of mediating computer system places it within the general tradition of *intelligence amplification* (IA).   This perspective takes the position that computer systems can be developed that partially mirror human mental functions; thus, by increasing the capacity or speed of operation of those functions, these systems can thereby increase or amplify the mental capacity of the human user working with them. As a result, quantitative increases in specific functions may produce qualitative differences in intellectual behavior, making the computer a necessary but not sufficient tool for enabling this mode of thinking.

Vannevar Bush (1945) is generally credited with originating the idea of intelligence amplification. Writing before the first commercial computers were developed, Bush described a hypothetical desk-like device he called the *memex* that would be implemented using microfilm technology. It would permit a human user to store vast quantities of data, add new information, but, most important, add cross-references at the bottom of any microfilm page that could be instantly followed to some other page. Thus, the human user could construct large networks of semantic relationships within the memex, drawing together vast quantities of data and then quickly and associatively move from one intellectual context to another. One could argue that the book — or at least a library of books — could similarly extend the capacity and precision of human long-term memory and that books do, in fact, include similar cross-references. Bush's innovation lay in the speed with which associative links could be followed to access new material — a second or two versus the minutes or even hours required to move from one printed volume to another.

It makes sense to talk about Bush's memex as an amplifying device in the following sense. He identified several key architectural features of human intelligence — long-term memory, semantic relationships, and associative access — and then provided within his memex — at least in theory — their operational counterparts, but with greater capacity (the microfilm store with its embedded semantic relationships) and comparable speed (associative access). Thus, Bush believed his device could amplify a specific set of human mental functions. No one has yet built a complete memex as Bush described the device. However, using more familiar computer technology, Doug Engelbart was the first to build a memex-like system (Engelbart, Watson, & Norton, 1973). In recognition of the goal to supplement human intelligence, Engelbart called one version of his system *Augment*. Today, many of the features first described by Bush and first built by Engelbart are routinely found in contemporary hypertext systems, some of which are discussed in chapter 3.

Just as IA systems make possible a type of mental behavior that would not be possible without them, so, I suggest, a particular type of collaboration support system may enable a form of collective mental behavior that would not be possible without it. These systems, I suspect, will be based on principles analogous to those for IA systems, but with important distinctions and extensions. We normally form collaborative groups for two reasons. First, the task is too large and/or there is not enough time for it to be done by one person. Second, no individual possesses all of the skills and/or knowledge required. However, when we (necessarily) assemble a group to overcome these problems, we inherently create other problems. Because the intellectual construct being developed by the group is likely to be too large to be known in its entirety by any one individual, it may lack intellectual integrity. Rather than being a structure that is deeply principled and elegantly simple — as we expect of the work of our best individual minds — it may emerge as an awkward assembly of incongruous pieces. Indeed, we have come to expect this of groups, as indicated by the old joke that a camel is a horse produced by a committee.

A computer system that can help groups approximate a CI will have to include, as a minimum, functions that help them perceive and address the overall structure and integrity of their work. It must include tools to help groups establish and maintain the internal consistency and coherence among the various information products they produce through the individual hands of their various members. Thus, it will have to amplify intellectual skills that are (relatively) strong in individuals but less so within groups. It may also include additional tools to facilitate access and version control, communication, joint work, and other group behaviors. But it cannot neglect the more basic requirements of intellectual integrity, coherence, and consistency.

Overview

The approach I take in building a concept of collective intelligence is to consider collaboration as a type of information processing activity. Thus, I look at several Information Processing System (IPS) models and architectures of individual cognition, identify key components and functions within them, and then identify constructs within collaborative groups that are recognizable as extrapolations of these components and functions. I should point out that there is no inherent reason to believe that a collective intelligence should necessarily resemble familiar models of individual cognition; it could have an entirely different structure. But, if we can see a resemblance between the construct identified as CI and commonly accepted models of human cognition, to which we attribute intelligence, then we are likely to be willing to attribute intelligence to that construct, as well. On the other hand, if the structure identified as CI were entirely different, it would require more extensive justification to extend the claim of intelligence to it.

The volume is dividend into two parts. Part I discusses foundation concepts that are used in Part II to build a concept of collective intelligence and to inform that discussion. Chapter 2 considers the range of activities found in collaborative groups as a result of differences in size, scale of work, task domain, and so forth, by considering three different collaboration scenarios. A simple model of basic information types and the flow of information as one type is transformed into another is also presented. Chapter 3 discusses computer support for collaboration. It reviews a range of system features that fall within the general category of CSCW systems and identifies key features needed to develop the different information types noted in chapter 2. It also describes one particular system in more detail that serves as the reference system for the rest of the discussion. Chapter 4 discusses IPS models and architectures in order to identify key components. It describes both general models/architectures as well as specific IPS models for particular tasks and particular circumstances (i.e., human–computer interaction).

Part II tries to build a concept of collective intelligence. Chapter 5 discusses the different memory systems found in computer-supported collaborative groups that can be recognized as extrapolations of human memory systems. These constructs function as a form of *collective memory* for collaborative groups. Chapter 6 identifies several types of conceptual processing found in collaborative groups that are analogous to individual conceptual processing. They form a concept of *collective processing*. Chapter 7 considers issues of *collective strategy* within large multigroup collaborative projects. Chapter 8 examines issues of *collective awareness* and *control*. Chapter 9 concludes the discussion by identifying a set of research dimensions that could provide a framework through which to view and relate a broad range of research and development in the field. It also looks briefly at implications for a theory of collective intelligence.

Part I

Foundation Concepts

In Part I, I lay the foundation on which a concept of collective intelligence is built in Part II. Because the form of collaboration being considered is both intellectual and computer-assisted, essential concerns include the types of information groups produce, the tools they use, and their social and conceptual behavior as it relates to knowledge-construction. Each of these concerns is addressed in Part I.

First, I look at behaviors found in typical collaborative projects that range from several weeks to several years in duration. A common thread that runs through them is their processing of different types of information and their transforming of one type of information into another. Thus, viewing collaborative work from an information processing perspective seems natural and straight-forward. To aid in characterizing collaborative behavior from this perspective, a model of information types and flow is described.

Second, I discuss computer and communications tools for collaboration that help groups produce the types of information identified in chapter 2. The discussion includes systems that support the independent, asynchronous work of a group's individual members as well as their synchronous, collective work. One particular system that provides both forms of support is described in more detail and serves as the example system in Part II.

Third, I review major information processing system (IPS) models and architectures for cognition. The discussion begins with general models, including Newell and Simon's original IPS model, Anderson's ACT*, and Newell's SOAR architecture. It also includes situated models, including an IPS model of human-computer interaction and specialized models for specific tasks, such as writing. A set of basic architectural components is identified, including long-term memory, working memory, processing operations, and problem-solving and knowledge-construction strategies. Each of these components or functions is examined in Part II with respect to collaboration.

The materials discussed in these three chapters provide a basis from which to consider collaboration as a form of computer-supported information processing behavior.

Chapter 2

Collaboration as an
Information Processing Activity

In most intellectual collaborations, the work of the group is ultimately focused in the production of some concrete product that constitutes successful completion of the project. Although groups differ significantly in their size, duration, the complexity of the products they produce, and the ways they go about doing their work, they also show surprising similarities in many of their activities, such as developing a body of shared knowledge about the task, using that knowledge to develop a plan of action, and, in turn, using the plan to guide their work. In this chapter, I illustrate both the diversity among collaborative groups as well as their similarity with respect to the types of information they use and produce.

The chapter is divided into two sections. First, three different collaborations are described to illustrate the range of behaviors commonly found in collaborative groups and to ground the discussion in concrete detail. Second, a simple information flow model is discussed; it describes the flow of information from one type to another, as concepts expressed in one form are transformed into a different form.

Three Scenarios

The scenarios that follow describe three collaborative projects that range in duration from several weeks to several years. The groups range in size from 4 members to as many as 20. And their work results in an equally diverse set of products. In the first, an ad hoc group representing three different employment categories in a university academic department work out a departmental policy for assigning parking permits. In the second, a task group in a federal agency is charged with planning and preparing the testimony a representative of that agency will give before a Congressional committee. The third scenario describes a larger group working in industry to develop a new software system along with its supporting documents. Although each scenario originated with an actual group or situation, I have taken liberties in the discussion to illustrate particular points. Thus, they should not be read as accurate descriptions of actual groups.

Scenario 1

The first scenario concerns an ad hoc group formed within an academic department in a large university. Their charge is to draft a proposed departmental parking policy. The issue has recently become sensitive because of the closing of a large, nearby lot to make room for a new building. The group is composed of four members: representatives of the secretarial and technical support staff, the tenured or tenure-track teaching faculty, and the research faculty holding fixed-term appointments supported by outside research contracts, plus the associate department head who serves as chairperson of the group. They have 3 weeks in which to solve the problem, draft the policy statement, and report back to the head.

The group met right away to begin work. It quickly became apparent that parking was an emotional as well as practical issue for some of the members. The representative of the support staff told the group that his constituents thought spaces should be assigned solely on the basis of seniority. Because many of the staff had worked in the department for more than 10 years, this would mean that no desirable places would be available for new faculty to be recruited over the next few years, a position that was said to be unacceptable to the department head. The representative of the regular faculty argued for using seniority within employment category as the basis, with the faculty, research faculty, and staff categories falling in that order of priority. This was essentially the status quo. The research faculty representative proposed a rather complex formula that computed a number for each person based on points assigned for job category, rank within category, seniority, and several other factors. Although the group remained relatively civilized toward one another, the first meeting ended with no solution in sight.

Over the next week and a half, the group met several times. At times, the meetings seemed productive; at other times, tempers flared. But, gradually, the group settled down and began to analyze the problem. Between two meetings, one member analyzed the data regarding lot requested versus lot assigned based on the current assignment scheme and concluded that the issue really boiled down to a shortage of eight "reasonably attractive" spaces that were desired by someone, but not available. Attempting to be funny, another member suggested that the department should rent those people spaces in a nearby commercial lot. After a few more wisecracks, someone asked the question: "Why not?" The group all looked at one another, paused, and then began to list the reasons why this could not be done. It would cost too much money. It would still mark some people as "second class citizens." And so on. But for each such reason, someone suggested a possible way around the problem. By the end of the meeting, they had not reached consensus, but no one seemed completely sure that the commercial lot option was impossible.

For the next few days, small knots of people could be seen talking earnestly among themselves, and one had the impression that most of the department was talking about the problem. When the team met several days later, the associate head reported that she had run a spread sheet on the costs and found that the difference between what the university charged for a parking space and what the commercial lot charged was only $15 a month. Because the total came to less than $1,500 for the eight spots per year, she felt it was financially feasible. Her remark was interpreted by the group to mean that she had also discussed this option with the department head and that he was willing to find the necessary funds. The staff representative still objected to the status implications, but he was having increasing difficulty arguing this position. Sensing this, he finally agreed to go along with the commercial lot option if the department would upgrade the workstations of people voluntarily accepting a less attractive lot assignment than they were entitled to. Because the department planned to upgrade all workstations over the next 2 years, the associate head said she saw no reason volunteers could not be given priority. With this, the group sensed that they had an agreement that, if not entirely to everyone's satisfaction, was at least workable.

While two members chatted, the other two drafted a short three-paragraph report, which they all then briefly discussed. After making several minor changes, the group asked the chairperson to produce a clean copy of the report and submit it on their behalf to the department head.

In this scenario, the product produced by the group consisted of a one-page document written at the very end of the process. However, the size of this document does not accurately reflect the work of the group. A lot of information flowed back and forth within the group during their discussions. It also flowed between the group and the other members of the department in the form of both informal conversations and more "official" contacts such as that between the group chairperson and the department head. Most of this information remained intangible — in the heads of the participants — but it was crucial to the success of the group. The matter of upgrading workstations, which was not really related to the parking issue, addressed a social, rather than a conceptual, aspect of the problem. But, it, too, was crucial in making the solution palatable to an important constituency of the department.

Thus, the behavior of the group must be understood in both social and conceptual terms. Social aspects included both behaviors within actual group meetings as well as several different forms of interaction between the group and the rest of the department. Conceptual aspects included both tangible and intangible forms of information. The amount of intangible information generated over the 3 weeks was large compared with the tangible product finally produced by the group — the policy statement that was their charge. But, clearly, both forms of information were needed, and, for this task, probably in this proportion.

Scenario 2

An investigative agency of the federal government has been asked to testify before a Congressional committee on potential use of the military in domestic drug enforcement. The agency has standing working groups concerned with the military, drug-related issues, and legal issues — each in a different division of the agency — but it has no established group concerned with this particular combination of issues. As a result, a five-person team — comprised of three area specialists, the person who will testify as the representative of the agency, and her chief assistant — is assembled to plan and develop the testimony. The group is responsible for writing the approximately 15-page statement that will be entered into the official record of the committee and any visual aids or supporting materials that will be used by the witness during her testimony. Just before the hearing is held, the witness will plan her own brief (typically, 5-minute) oral presentation, based on the written statement developed by the group. The witness is the team leader and has authority over the work of the group with respect to this project. The group has approximately 2 1/2 months in which to complete the task. Although the assignment is regarded as important and high-profile, team members all have other responsibilities and assignments within their respective divisions that will also require their attention throughout the project.

The group began by reviewing previous work done in each of the three areas involved as well as information available from outside the agency. This task was divided among the members, following the first organizational meeting of the group, and occupied them for the first 2 weeks of the project. After that, the group met as a whole two or three times a week over the next several weeks. During these discussions, key points in the source materials that had been gathered were identified that might be relevant to the testimony. The group also compared notes on what they knew about the members of Congress on the committee and their known concerns and idiosyncrasies. Group members took turns listing points on a whiteboard, drawing various conceptual diagrams, and so forth. They also took turns writing up notes of their discussions and distributing them to the others.

Fairly early, the group crafted a one-sentence "message" to serve as the focal point of the testimony. This turned out to be a surprisingly difficult task. They could all identify important issues, but seeing which larger point they all added up to proved to be harder than they had anticipated. The message was also revised several times over the course of the project.

By the beginning of the fifth week, the group felt it was ready to shift gears and begin planning the actual written statement. During a particularly long meeting, they hammered out an outline that included the message, five major sections for the testimony statement, and the main points to be made in each. At the end of the meeting, they divided the sections up among themselves; most were assigned to a single individual, but two sections were assigned to two-person teams.

Throughout these discussions — and, in fact, throughout the project — the individual area specialists on the team met informally with their respective deputy division directors to keep them informed regarding the direction the project was taking. Occasionally, they would receive instructions regarding positions or arguments on an issue that were regarded as sensitive by the divisions. Although members would occasionally report these conversations, more often these concerns were simply absorbed into the perspective of the particular team member and voiced during the group's discussions. In general, the divisions made relatively few attempts to influence the message and strategy the group crafted, although they expected to be kept informed.

Over the next week, each individual or team was responsible for fleshing out additional levels of detail for the plan by identifying the specific facts or policy positions to be discussed under each point. At the end of that time, they again met as a group. During this meeting, they put their various pieces together, reviewed the whole extended plan, and revised it; they also assigned the task of looking further at several particular issues to the individual or team involved. At the end of the meeting, writing assignments were decided. One person was also given the task of collecting the final revised pieces of the plan and distributing copies of the whole to the group. Thus, by the beginning of the sixth week, the group was ready to begin drafting in earnest.

As group members wrote their individual sections, they referred to the overall plan for context — for example, one person discovered that an issue that had been viewed as a matter of policy was actually a point of law. Because legal issues were described in an earlier section being written by another member, this writer decided to meet with the other person to negotiate a change in the plan. The other writer agreed, and they sent a note to the rest of the group telling them of the change in the document plan. Several similar changes were negotiated by other members over the course of the project.

When drafts of all of the sections were complete, some 2 weeks later, they were assembled into a single document, and copies of the whole were sent to all five team members. Each reviewed the document alone before they met as a group to discuss it. It was then 3 weeks before the scheduled date of the hearing, and the document was rough. A lot of material was repeated, and the writing styles of several members were noticeably distinct. After some discussion of the whole document, the group went through it section by section, discussing potential problems. After two rather long meetings, they completed their detailed review and had marked a number of places for further work.   Again, the individual members of the group worked alone to revise their sections in accord with the discussion and the marked-up document.   As the sections were revised, they were again circulated, marked, and returned to their respective authors.   A "final" group draft was ready 10 days before the hearing.

At this point, they met with someone from the agency's Congressional liaison office to discuss the actual hearing event and to identify questions expected from the various members of the committee.   After the liaison person left, they also discussed visual aids for the presentation.   They went back and forth on the issue, but finally decided to prepare two large (30 in. x 60 in.) poster boards to be placed on easels at the side of the hearing room during the testimony.   One succinctly stated the "message" and three supporting points; the other showed in graphic form the results of an econometric model that predicted different impacts on availability of illegal drugs that would result from different levels of interdiction by the military. The order for the poster boards was placed with the agency's graphics department.

Following this meeting, the witness who was also the team leader did a complete editing pass through the testimony statement for consistency and to make the language closer to her own writing/speaking style.   Two days later, that version was passed to the head of the agency for his comments and, they hoped, approval.   He liked it and had only minor changes, which were made.   The statement was ready a full 48 hours before the hearing!

This collaboration is quite different from the first one.   Part of this difference can be accounted for in the size of the products developed.   Because the document that represented the goal of the project was longer, the group produced a number of intermediate products that were never intended to become part of the final document but served important instrumental roles.   These included the notes and diagrams on the whiteboard, notes of meetings, and the different versions of the plan for the document.   The intended product — the testimony statement — also went through several versions. Thus, the Scenario 2 task was more of a knowledge-construction task than a problem-solving task, as was the case in Scenario 1.

Although a great deal of important work went on in group meetings, the majority of time was spent in individual work —

examining resource materials, building and extending portions of the plan, drafting sections of the testimony statement, reviewing and revising drafts, and so forth. Although the group had to consciously attend to issues of internal consistency within the statement, this was not a particularly difficult or time-consuming problem for a document of this size. On the other hand, defining the overall message, particularly one that represented the consensus of the group, proved more difficult than they had anticipated. Finally, this group was more intellectual and less social in its emphasis, although like Scenario 1, Scenario 2 team members kept their constituencies informed as to the approach the group was taking. But, overall, the work of this group was far less emotionally charged than was the case for the group described earlier.

Scenario 3

The third scenario concerns a group working within a software development department in a large computer company. The group, which ranged from 15 to 20 people during the project, collaborated over a 2-year period to develop a system intended to be used within the corporation to support a new "quality management" initiative that was to begin soon. However, if the initiative proved successful, the company planned to market system, training, and consulting services to other organizations starting similar programs. Because this project is too large to describe in as much detail as the first two scenarios, I focus on ways in which it was similar and dissimilar to the other two.

For most of the project, the members of the group were divided into three technical teams responsible for developing the three main components of the system — the underlying database, the user interface and its function, and communications. These teams ranged in size from three to six members. The group also included a project leader, assistant project leader, and two clerical support staff. The project leader designated one senior person in each team as team leader; these same three staff members plus the assistant project leader also functioned as a technical advisory committee to the project leader.

Much of the group's work was concerned with writing several design documents that provided different perspectives on the system the group was developing; hence, the process they followed was often similar to that of Scenario 2, only repeated several times. Scenario 3 was different, however, in its greater reliance on diagrams and in producing an altogether different kind of "document" — the computer code that constituted the system, itself.

The (conventional) documents they wrote over the course of the project began with an initial "whitepaper" that provided a conceptual description of the system. It was written by the team leader before the project began. It also described the rationale behind the quality management initiative and listed a number of information management and access functions that would be needed if the initiative was to be successful.

This list served as the starting point for a second document — a more formal requirements statement that identified the specific computer operations the new system would support as well as other technical requirements for hardware, performance, the programming languages and toolkits to be used, and so on. The requirements document was written collaboratively by the project leader, the assistant leader, and representatives from several departments expected to use the new system.

A third document described the abstract design or architecture of the system. It was developed iteratively over a 3-month period. Overall design for the system was discussed intensely for the first few weeks in meetings attended by representatives from all three teams, then off and on for the next several months, and then, for the remainder of the project, only occasionally to resolve problems that arose. Once the general architecture became relatively firm — after the first month — the more detailed designs for the different parts of the system were developed by the individual teams. As the various segments were completed, they were assembled into a single document. Throughout this process, the project leader's technical advisory committee met to review the individual parts as they became available.

A more detailed specification of functions and interfaces supplemented the architecture document. Work on this document overlapped work on the architecture. These descriptions exhaustively defined the interfaces for each module as well as the functions included within them. As the architecture document neared "completion" and work on it tapered off, work on the specifications increased. However, both the architecture and the specifications remained incomplete throughout the project. As the system was coded and tested, the teams often found that changes were needed in the system design.   Consequently, they updated the architecture, specifications and, sometimes, the requirements documents throughout the project. A larger problem was created, however, when they did not make these changes.   This problem arose primarily from programmers making design changes in the system code but not updating the design documents to reflect those changes. Thus, toward the end of the project, the specifications and the other design documents became less and less consistent with one another and with the computer code that comprised the actual system.

A separate document outlined a set of tests that were used during development to test individual components and, near the end, to test the system as a whole.   One member from each team was given responsibility for writing the test plan for his or her team's part of the system. They, along with the assistant project leader, also functioned as a test group, particularly toward the end of the project.

The last major document was a set of user instructions. An initial version was actually written quite early in the project as part of the specification for the user interface, but it was barely more than an outline of user operations.   Near the end, two members of the interface group were asked to rewrite this document to include more rationale and explanation.   It took them approximately 2 months to produce a new draft; however, the final user documentation was not completed until well after the development project ended and then only after a technical writer rewrote the draft from scratch.

Thus, the project produced a number of "formal" documents that included the concept paper, requirements, architecture, specifications, test plan, and user documentation.   They also produced a number of "informal" documents; these included notes of meetings, memos and letters, various project plans and time-lines, progress reports, a budget, and brief explanations included in the computer program itself. More ephemeral were jottings on whiteboards, personal notes, and agreements and understandings that were reached but never documented. In general, the project took pains to produce and keep track of their formal documents, less so for their informal ones, and not at all for ephemeral products.   As a result, people would occasionally search frantically for some lost piece of paper, but, in general, the group got along relatively well with the materials they kept, aside from the problems of consistency noted previously.

Throughout the project, diagrams played a more substantial role in the group's work than in the other two scenarios. Drawings of various sorts comprised a relatively large part of the planning documents, particularly the architectural description. They were also prevalent in work at the whiteboard during meetings. The more abstract their discussions, the more important they became, particularly for ironing out differences in understanding of technical points and for exploring different design options. In fact, key decisions often turned on these diagrams, some of which were preserved in meeting notes but many of which were retained only in the heads of those attending the meeting.

The third major type of product produced by this collaborative group was the computer source code. Written in a programming language, this "document" constituted the actual goal of the project — the computer system the group was building. Consequently, all of the other products described previously were secondary, although writing the code without them would have been impossible. Just as the group produced several different kinds of documents, so they produced several different kinds of computer code.

Prior to the beginning of the project, the project leader and the vice-president in whose division the project was situated had decided that the system would be developed using a relatively new approach to software design known as object-oriented programming. This approach was said by its proponents to result in systems that are more modular, easier to understand and maintain, and that future projects should be able to reuse portions of their code with relatively few changes. In picking the project personnel, the project leader was careful to select several senior programmers experienced in both object-oriented design and programming, and it was from this group that the three team leaders were named. However, most of the other programmers were skilled in more traditional methods. As a result, the project faced a substantial learning curve at the beginning to absorb the object-oriented paradigm and to get up to speed in the specific object-oriented programming language selected.

Training was supervised by the assistant project leader, but most actual instruction was done by the three team leaders. For the first several weeks, all three teams met as a group and were instructed by the team leaders, taking turns. After that, training merged with other activities and was done on a team-by-team basis. During both stages of training, the individual group members wrote a considerable amount of computer code that was never intended to be part of the system. At first, they wrote small bits of code in order to understand how a particular technical feature in the programming language worked. After that, they wrote short programs to which they applied principles of object-oriented design. Later, they wrote code as a way of exploring and learning new toolkits and utilities. To some extent, exploratory programming continued throughout the project, but most of it occurred during the first 2 or 3 months and overlapped with early design.

The second type of computer code written by the project was the system itself. The project's general strategy was, first, to prepare a detailed design for a module as described previously and, then, to write the code. However, as development moved to more and more detailed levels of the architecture, designing and coding often merged. For example, early design did not always anticipate unusual exception conditions. When the teams became aware of issues such as these, they were forced to design and code at the same time, sometimes leading to inconsistencies between system design as documented and system design as implemented in the code. Once written, the code went through numerous "revisions" during debugging and testing. Gradually, the teams accumulated libraries of system modules that had been tested and were thought to be relatively free of errors. These were ultimately assembled to form the complete system. However, as problems were discovered and corrected, these libraries had to be updated accordingly and the system reassembled.

The third type of code produced by the project was code intended to assist with debugging and testing. Each team developed a set of test programs to simulate the functions of the modules their components interacted with, to provide an exhaustive set of test conditions, and to simulate different load conditions for evaluating performance. Although many of the test programs were written by the individual team members to test their own work, quite a few were eventually gathered into standard "test suites" used in the latter part of the project by the testing team to ensure that corrections made to fix one problem did not introduce new ones.

Several differences stand out in this scenario. Perhaps the most prominent one is the difference in scale. Scenario 3 was an order of magnitude longer in duration than Scenario 2, but it was more than two orders of magnitude larger in the total size of the products it developed. Whereas both Scenario 1 and Scenario 2 each went through one complete cycle of development for their respective documents, Scenario 3 went through many such cycles for the different documents and modules of computer code developed during the project.

These differences in scale also led to differences in project organization. Most of Scenario 3's work was done in relatively independent subgroups — thus, the project was a group of groups. In contrast, Scenario 2 functioned largely as a single coherent group that periodically delegated tasks to individuals or two-person teams but quickly assembled their independent efforts into a whole. Thus, one could normally say that at a given time the group as a whole was engaged in a particular activity, such as brainstorming, planning, drafting, editing, and so forth. In contrast, the different Scenario 3 teams progressed at different rates through their respective work plans and merged their independent efforts far less frequently. Consequently, one could not say at any given time exactly which activity the group as a whole was engaged in, because they were often doing different things.

Model of Information Type and Flow

The three projects described previously are quite different from one another in terms of scale, duration, task, and the intellectual activities they used to accomplish their respective goals. But if we focus on the types of information they used and produced, we can also see similarities.

Each of the three groups was concerned with producing some type of tangible product that represented successful completion of the group's primary goal. In Scenario 1, this target product was the brief written recommendation regarding parking policy they wrote during their final meeting. In Scenario 2, it was the 15-page written testimony statement that was included in the record of the Congressional hearing and the two supporting poster boards referred to by the witness in her oral statement. In Scenario 3, the target product was the computer system defined by the source code and the supporting documents, such as the user documentation, delivered with that system.

However, target products tell only part of the story. The groups also produced instrumental products to help them with their task.

These ranged from meeting notes and whiteboard jottings, to plans and outlines for the target product, to large independent documents, such as requirements and test plans. Some of these, such as the whiteboard drawings, were ephemeral. They came into existence for a brief time, served their purpose, but were then destroyed or lost. Others, such as the software design documents, were persistent; they were kept by the group and maintained with varying degrees of consistency from the time they were created until the end of the project.

In addition to these tangible forms of information, all three groups also used and/or produced several types of intangible knowledge. Intangible knowledge is carried in the heads of group members. In some cases, it was eventually incorporated into one or more tangible products. For example, in Scenario 1, the organizational change that was finally recommended was first discussed, along with several other possibilities, both inside and outside the group before it was finally encoded into written form. Thus, the information existed in intangible form before it was transformed into a tangible, written document. In other cases, information remained intangible throughout the project. This was true for the options discussed but not selected by the group in Scenario 1.

In some cases, intangible knowledge is shared by the group as a whole. For example, during early group discussion, individual members often voice their different understandings of the group's goal before the group eventually settles on a common approximation acceptable and known to all members. However, not all shared knowledge is developed through explicit group activities; for example, shared knowledge is also carried in the culture of the organization. In Scenario 2, all members of the group knew the general format of a Congressional hearing; what was not common knowledge were the idiosyncrasies of the individual Congress members on the committee.

In other cases, intangible knowledge remains limited to a specific individual or a subset of individuals and, thus, is private with respect to the group as a whole. This is frequently the case when members are selected because they have particular expertise or perspectives. During discussions, some of an individual's private knowledge may become shared as that person speaks to issues from his or her base of private knowledge, but much of this base remains out of sight from the group as a whole and comes into play only indirectly by informing that individual's work or views.

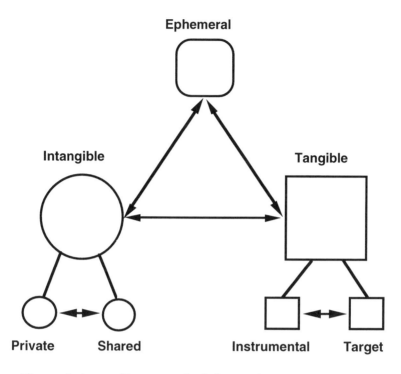

Fig. 2.1. Types of information produced by collaborative groups and the transformations from one type to another.

Thus, collaborative groups as a matter of course develop three basic types of information: *tangible, intangible,* and *ephemeral.* Tangible knowledge can be divided into *target products* that represent successful completion of the group's task and *instrumental products* that support the group's work on the target product but are not part of that product. The collection of target and instrumental products developed and maintained by a group during a project constitutes the group's *artifact.* Intangible knowledge does not take tangible form but, rather, remains within the heads of the members of the group. Some is *shared* to an approximation by all members of the group; other is *private* with respect to an individual or a subset of the group. *Ephemeral products* lie somewhere between tangible and intangible knowledge. This information is given physical form for brief periods of time, but unlike the instrumental products that are included within the artifact, ephemeral products are destroyed or lost. These three basic types of information and their respective subtypes are shown in Fig. 2.1.

The collaborative process produces a *flow of information*, as information in one form is transformed into information in a different form. This flow is indicated in Fig. 2.1 by arrows between information types. For example, during group discussions, private knowledge held by one member may become shared knowledge held by the group if that person explains a privately held concept to the group. When this happens, information can be viewed as flowing from one individual to the group. If the individual with the private knowledge uses the whiteboard to draw a diagram to help explain the concept, the information is transformed from private to ephemeral to shared knowledge; Fig. 2.2 shows this particular flow. There are a number of different flows that can be observed in groups, produced by a variety of different processes and activities. It remains a task for research to uncover these.

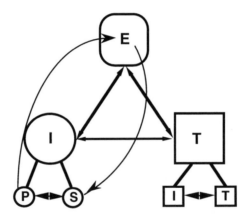

Fig. 2.2.   Flow of information from private to ephemeral to shared.

Collaborative groups produce a mix of products and knowledge, but different groups go about it in different ways, on different scales, in accord with different strategies. Consequently, groups can be characterized according to the nature of the artifact and the flow of information from one type to another they produce over the course of a project. To see this, let's look back briefly at the three scenarios from the perspective of the information flow model.

In the first scenario, most of the group's time and effort was spent building a body of shared intangible knowledge, that is, analyzing the parking problem from the perspectives of the different constituencies and finding a solution acceptable to both the members of the group as well as those they represented. The actual target product consisted of only a short three-paragraph statement written at the very end of the project. Ephemeral products were also minimal and the group produced no instrumental products.

In the second scenario, the target product — the 15-page testimony statement and the two accompanying poster boards — was much more substantial, and direct work on it constituted a much larger proportion of the group's effort. This group also produced several different instrumental products — for example, meeting notes and the statement plan — that played important roles in the project. Scenario 2 undoubtedly developed a much larger body of shared knowledge than did the Scenario 1 group; however, this difference is unlikely to be proportional to the difference in size of their respective target products, which had an approximately 20:1 ratio. On the other hand, private knowledge played a much more important role in Scenario 2, because individuals and two-person teams, drawing largely on their individual areas of expertise, developed the detailed plans for the different sections of the statement and then wrote those sections.

The Scenario 2 group also went through a number of shifts between private and group work. This strategy permitted work to go on in parallel; but it also created problems of consistency, for example, when the group integrated the different parts of their detailed plan and, later, when they merged the draft sections of the testimony statement. However, because the scale of the project was relatively modest, resolving these inconsistencies was not particularly difficult or time-consuming.

The Scenario 3 project was several orders of magnitude larger and more complex than the Scenario 2 project. Its target product — the computer system — was a large "document" consisting of more than 1,000 pages of source code and accompanying comments. Instrumental products included a number of complete documents. The body of intangible knowledge shared by the group as a whole included knowledge of project goals, the large-scale architecture of the system, basic principles of object-oriented design, and a new programming language most team members learned at the beginning of the project. Although the volume of shared knowledge for this project was larger than that of the other two, it was probably not proportionally larger relative to the difference in size of their respective artifacts. On the other hand, private intangible knowledge played a much larger role in Scenario 3 than in the other two scenarios, because individual teams possessed and developed specific technical expertise in their assigned portions of the system design. Finally, ephemeral products played a larger and more important role, particularly during the various design activities of the group.

Consistency requirements were more stringent and more apparent for the software development group. In Scenario 2, consistency relationships existed primarily within a single document, the testimony statement; however, in Scenario 3, consistency relationships also existed between documents, such as between the architecture document and the source code. Because of this and because different components were developed by different teams, keeping the whole artifact consistent was much harder. In fact, the group did not always do this, particularly toward the end of the project. These inconsistencies caused problems for the development group, and they can be expected to cause problems for those who will maintain the system in future years. Thus, as a project grows in size and complexity, maintaining consistency and coherence in the artifact is likely to become increasingly important and increasingly difficult, requiring more and more of a group's resources.

Issues for Research

Implicit in this discussion are a number of unanswered questions and issues for further research. In this section, I briefly discuss several of these. This list, of course, is not exhaustive. Rather, it is intended to suggest the type of questions and issues that can be inferred from the discussion. I hope that others occurred to the reader, that they were noted, and that they will stimulate discussion and, perhaps, future research.

- *Develop detailed portraits of specific collaborative groups*

Detailed descriptions for a broad range of collaborative groups would be valuable for a number of reasons. First, they would open the area up for inquiry so that we could see what collaboration is as a general mode of work across different tasks. From this data we could also infer a repertoire of both common and unusual behaviors. And we could begin to see which behaviors and patterns of work tend to result in more effective or more efficient collaborations versus those that are less productive. Latour & Woolgar (1979) provided a good example of close description based on ethnographic methods; for computer-based collaborations, we should be able to supplement their methods with system recorded data to observe a greater range of individual and collective behaviors as well as the rhythms and interactions between the two.

- *Test and refine the information flow model*

The information flow model needs to be validated against a wide range of collaborations in different organizational contexts. As detailed portraits of groups become available, the model should be tested against those data and updated as needed.

- *Identify the vocabulary of transformation processes that occur in collaborative groups*

A related issue is further refinement of the model. Each of the transformations depicted as a line between information types signifies a wide range of more detailed processes. For example, the transformation from private to shared intangible knowledge frequently takes place in group meetings (it is not, of course, limited to such gatherings), but it occurs in different ways. It can come about through a statement made by an individual in a discussion, a statement within a more formal presentation, or a response to a question. Each such action consists of a complex structure of finer-grained social and intellectual processes. Thus, by a vocabulary of collaborative actions I mean a list of the different kinds of behaviors that transform information of one type into information of another type. Such a list will, of course, be extensive if it is to cover the many different situations in which collaboration takes place. But identifying them will be worth the effort because these actions could serve as the basic elements in a process model of collaboration.

- *Identify sequences of collaborative actions*

Collaborative actions normally take place in sequences or "phrases" rather than as isolated events. For example, a discussion normally includes a number of different statements by different members of the group, interspersed with individuals' drawing diagrams, referring to documents present or absent, showing transparencies, waving their arms in the air, and so on. These different events are woven into a fabric of discourse. If identifying a vocabulary of actions is a first level of analysis and characterization, identifying short repeated sequences of actions is a second level. Groups use such sequences to build small component structures and/or to link new information into an existing conceptual structure. Thus, if we are to understand how groups develop large, complex artifacts over long durations, we must first identify the basic process sequences they habitually use.

Chapter 3

Computer Support
for
Collaboration

In the preceding chapter, I described an information flow model that identifies several types of information produced by groups. The model provides a lens that lets us penetrate the task domain and see more deeply into the fundamental behavior of groups as they go about the task of building large, complex structures of ideas. As a result, we can characterize groups in terms of generic behaviors that produce those types or transform one type into another. These behaviors are part of a general process of collaboration.

In carrying out this process, groups use a variety of tools. Before the computer, they used paper and pencil, blackboards, and the postal system. After computer and communications systems became widespread, groups adapted these new resources to their collaboration needs — by taking turns editing the same file, by mailing diskettes to one another, and by exchanging files as well as messages through e-mail. Recently, a new generation of tools has begun to appear, designed from the start to support collaboration.

These tools can all be related to the information flow model. Some are used to construct the artifact, others help transform private knowledge into shared knowledge, still others are used to produce ephemeral products. Inevitably, these tools affect the thinking of the individuals and groups that use them. Consequently, the process of collaborative knowledge-construction should be considered in the context of the tools groups use.

Collaborative software tends to fall into one of two large categories: *synchronous* and *asynchronous* tools. Synchronous tools support the simultaneous interactions of two or more members of a group. Shared editors and video conferencing systems are examples.

Asynchronous tools support the independent work of the various members of a group working alone. Electronic mail and distributed file systems are examples of the second category.

The goal for this chapter is to provide a basic understanding of collaboration technology that can be drawn on during the rest of this discussion of *computer-based* collaboration. Because this technology is changing quickly, the discussion emphasizes services provided by categories of systems rather than individual systems, although brief descriptions of selected systems are included to sharpen the image of the class. Thus, it is not intended to be a comprehensive survey of collaboration technology. Near the end of the chapter, one particular system is described in more detail. It combines many of the features found in separate tools and serves as the reference system for the rest of the discussion. The chapter concludes by looking briefly at several issues for research in collaboration systems.

Asynchronous Tools

Asynchronous tools enable collaboration by helping the members of a group work independently on the same tangible product, such as a document, a set of architectural drawings, or the source code for a computer system. Some — such as e-mail, file transfer protocol (ftp), and distributed file systems — are part of the larger system that we routinely work with, and we may not think of them as "collaboration tools." Others were developed from the start with collaborative use in mind.

Virtually all collaboration tools — synchronous as well as asynchronous — depend on some form of underlying computer network to provide communication between workstations or between workstation and specialized processors, such as file servers and supercomputers. Although there are a number of different networks, the Internet has emerged as the de facto standard and the communications infrastructure most collaborative users will have access to. A vast, amorphous federation of separate but interconnected networks, the Internet has become the web that connects the world — or at least those people who use their computers to connect to a network for non proprietary purposes. This includes researchers, educators, government personnel, people in the military who do not require secure channels, as well as individuals who work for many commercial organizations. It excludes specialized services, such as automatic teller machines, secure military channels, and private networks maintained by individual corporations, such as IBM and Walmart. Consequently, in discussing tools that rely on an underlying computer network, I do so with reference to the Internet. The discussion here covers only a small subset of services; a number of introductions and guides to the Internet describe other services available through it (e.g., Krol, 1992).

E-Mail

By far the most successful collaboration tool is electronic mail. E-mail permits a user connected to a computer network to send a message to another user also connected to the network. One normally sends mail by informing a mail application program of the recipient's name or system ID and his or her network address. The messages one receives accumulate in a file or "mailbox" on his or her local system, which can then be read using a mail application program.

Messages are generally formatted like a memorandum, with header information at the top indicating sender, receiver, date, subject, and so forth, followed by the text of the message. The message can be typed directly, or the sender can also copy a computer file into the body of the message. Thus, collaborators can exchange comments on a document they are working on, or, using the file copy option, they can exchange entire versions. Of course, files for other types of information in addition to text can be sent.

Although different networks use different address formats, most now recognize the form supported by the Internet. Because many of these same networks are physically linked to one another, through so-called *gateways*, it is now possible to send e-mail to someone located virtually anywhere in the world, have it relayed from network to network, and end up in that individual's local mailbox. Thus, it is almost as easy to send a message to a collaborator on another continent as it is to someone across the hall.

Although e-mail's potential as a tool for remote collaboration is obvious, the ways it has been used for that purpose have not always been what was expected. When e-mail was first introduced, many people thought that one of the strongest effects would be to facilitate collaborative relationships among people located at a distance. Indeed, this effect has been observed but, in some cases, not as strongly as expected. In a recent study done by the RAND Corporation, researchers found that people rarely sent messages to people at distant locations, and nearly half of the messages sent were to people in the sender's immediate vicinity (Bikson & Eveland, 1990). The more important effect, they speculate, may be overcoming temporal, rather than spatial, barriers. Whereas individuals may not be available for a visit or telephone call, they can accomplish a number of interactions over a day through sequences of e-mail messages and responses.

E-mail can also affect the ways groups organize themselves. In organizations where individuals participate in multiple work groups, e-mail tends to increase the number of teams an individual participates in (Bikson & Eveland, 1990). Similarly, it tends to spread leadership roles more evenly across the pool of participants. It also results in fewer face-to-face meetings, telephone calls, and other traditional forms of communication, although face-to-face meetings were preferred for resolving conflicts, reaching consensus, and addressing other complex and/or sensitive issues. There is also evidence that groups that make effective use of e-mail tend to be more successful than groups that do not (Fineholt, Sproull, & Kiesler, 1990).

FTP

Although e-mail can meet many of the communications requirements of collaborators, there are some needs for which it is not well-suited. For example, it is easy for someone to send a file he or she has been working on to a collaborator, but it is less straightforward for someone who wants a file located on another person's system to get that file. One can always send an e-mail message to the owner and have that person mail the file back. But that requires considerable human intervention. Many times, one would rather copy a remote file directly, without having to ask someone to send it.

This can be done using an Internet facility called ftp, an abbreviation for file transfer protocol. Whereas e-mail is sender driven, ftp is receiver driven. Ftp allows the person who wants a file stored on a remote machine to have that system send a copy of the file to his or her local machine, without involving another human being. This is done by starting an application program, called ftp, on the local machine and then contacting the machine where the desired file is located, through the Internet. Once a connection with the remote machine is established, a similar ftp program is started on that machine, as well. One can then send files back and forth between the two machines, subject to several constraints. The user must have the permission of the owner of the file to copy it, know the name and location of the desired file, and be a registered user on both systems.

To get around these restrictions, many sites have established anonymous ftp directories in which various public documents — such as technical reports, program documentation, and even computer programs — are placed. Anyone may access these files, and they are stored in the same logical location on all systems. A number of specialized collections of information are beginning to appear as anonymous ftp directories, ranging from aeronautics to zymurgy (Krol, 1992).

Thus, private ftp is well-suited for small groups of collaborators who wish to share a specific set of files, whereas anonymous ftp lends itself to less structured forms of cooperative work where the user population is larger and not necessarily known to one another.

Network File Systems

Although ftp and other forms of explicit file transfer are extremely useful, they still require considerable human overhead. Users must know the name and location of the system where the desired file is stored, invoke a file transfer program on both the local machine and the remote machines, and assume responsibility for maintaining consistency between the original and the copy, if that is a concern. These problems are solved and a number of additional services provided by network file systems (NFS). Examples include Sun's NFS (Walsh, 1985), DECnet (DEC, 1985), and the Andrew file system (AFS) (Spector & Kazar, 1991). Because AFS appears to be emerging as the de facto standard, at least for the UNIX and Internet worlds, and because it provides services not yet available in other systems, it serves as the example system for this class.

To understand how AFS and other similar systems can help groups work more easily with a common body of material, one must first understand the naming conventions used by various kinds of file system. Individual computers provide users with a hierarchy of directories (*folders*, in Apple terminology) and files. One can think of this hierarchy as a tree or organization chart, with the so-called root directory at the top. The root directory contains a set of other directories and/or data files. Each of these directories, in turn, can contain other directories and files, and so on. Because the directories and files have names, as well as the particular computer, one can identify any given item stored on that system by listing the names of all of the directories one must go through, from the top down, before reaching the desired item. For example, on my personal computer, this chapter of the book might be referred to as: */jbs/docs/CI/chapter3*, where *jbs* is the name of my computer, *docs* a directory of documents, *CI* the directory for this book, and *chapter3* the current chapter. The overall hierarchical structure of names is sometimes called a *name space*.

Network file systems extend the name space to a collection of machines. The first extrapolation is to machines located on the same local area network (LAN) or collection of interconnected LANs. Such configurations can be found in academic departments, corporate sites, or other similar geographic and administrative centers; they can contain hundreds of computers, including both individual workstations and specialized file servers. Network file systems such as AFS permit such collections to share a single logical hierarchy of files that integrates the files stored on all of the individual machines in the network. This name space is similar to that described previously, but has an additional tier added to the top. The root of the tree now refers to the site or collection, as opposed to the individual computer, whereas the second level names each machine in the collection. In this expanded name space, the example file might be referred to as */unc/jbs/docs/CI/chapter. unc* is the site, whereas the rest of the name remains the same.

This capability is obviously useful for collaborative groups. For example, a project can set up a system of directories and files that comprise the artifact, as discussed in chapter 2. Each member of the group could then view and edit this material, add and delete files, and so forth. Instead of having to consciously move files from one machine to another, as required by ftp, users can access files stored on any of the machines simply by referring to them by name. After that, the system does the rest. Of course, the system leaves it up to the group to keep up with logical relationships among files and directories (e.g., which files constitute the chapters of a book or the current version of a system), to maintain internal consistency within the data, to ensure that one member's changes do not undo the work of another, and so forth. Nevertheless, network file systems make it much easier for a group of collaborators to work with a shared collection of material.

A wide-area file system extends the name space one step further to include any machine or site that is accessible through the Internet and runs the wide-area file system in the local environment. This is done by adding still another tier to the top of the name space tree. In this expanded name space, the example file might be referred to as */afs/andrew.unc.edu/jbs/docs/CI/chapter3*. *afs* refers the entire structure — in this case, all of the various Andrew file systems located at different sites that have been drawn into a single logical structure — *andrew.unc.edu* is the Internet name for one particular site — unc — whereas the rest of the name below the site remains the same. Andrew then provides access to any file in this *entire* (potentially) worldwide system in exactly the same way it provides access to a file in the same local area. Thus, collaborators located anyplace in the world can work on a common set of files with virtually the same convenience they could were they located in the same building, so long as they have access to the Internet. The only differences are the longer file names and the additional time needed for access, caused by network delays and system overhead.

Collaborative   Applications

The systems described previously are parts of the basic computing infrastructure. They provide general services, such as exchanging messages and transferring or accessing files. Applications, on the other hand, provide functions that relate more directly to substantive aspects of the task at hand, such as a text editor being used to write a document or a drawing program for drawing a diagram. Most applications designed explicitly to support collaboration are synchronous, such as shared editors or meeting support systems. Three major classes of these tools are discussed in the section that follows. However, a few asynchronous applications have been built that support asynchronous work, primarily collaborative writing and/or document management. In this section, I briefly describe several of these applications to complete the discussion of asynchronous tools.

One particularly innovative application is a writing tool developed at Carnegie Mellon University, called the "work in PREParation" editor or, more commonly, PREP (Neuwirth, Kaufer, Chandhok, & Morris, 1990). It uses a two-dimensional grid, similar to a spread sheet, to enable the members of a group to work individually on a document they are co-authoring. The core text is assigned a column and is divided into horizontal segments, typically one paragraph per cell. Each participant is then assigned a different column in which he or she can comment on, rewrite, or otherwise edit the core text. Members can also comment on each other's comments, as well. Thus, one can scan horizontally across the display and see all of the comments and revisions for a given section of the text or scan vertically to see all comments and revisions made by a given individual. As might be expected, PREP has proved to be most useful for dealing with small-grain writing problems — such as style, sentence clarity, and paragraph structure — and less so for large-scale problems — such as overall organization and document structure.

Although PREP provides a structured environment with respect to the document and group members' comments and revisions, it does not impose structure with regard to members' social roles or authority. This design philosophy contrasts with that of Quilt, a collaborative writing system developed at Bell Labs (Leland, Fish, & Kraut, 1988). Quilt takes the opposite approach by encouraging groups to establish well-defined social and authority structures with respect to co-authored documents. The system includes three participant roles (co-author, commentator, and reader), six types of documents or comments (base document, suggested revision, public comment, private comment, message, and history), and a set of functions or actions users may employ (create, modify, delete, attach a suggested revision, attach a public comment, attach a private comment, attach a directed message, and read). Like PREP, Quilt enables users to attach comments to documents but through a form of hypertext link that must be followed by a reader to see the comment. One nice function found in Quilt that is not part of PREP is the capability to attach messages that invoke e-mail responses, such as a reminder to someone to look at a particular section of a document.

An innovative asynchronous application that is not a writing tool per se is Oval, an acronym for *objects, views, agents,* and *links,* developed by Tom Malone and his colleagues at MIT (Malone, Lai, &

Fry, 1992).   Oval is a meta-application that enables users to create their own specialized cooperative work applications.   They do this using Oval's four basic components.   Objects are collections of attributes that can be defined by the user, such as a form that includes names and addresses.   Views present particular representations of objects or collections of objects.   Agents are sets of rules that can operate on objects and the other components without the direct attention of the user.   They can perform a variety of actions, such as deleting objects older than a certain date or sending e-mail to the user to alert him or her of a change made to an object by someone else. Links provide a mechanism to build structures of objects.   Such structures can be used for a variety of purposes, from composing documents from collections of objects to providing the paths along which agents move to carry out their actions.   As a test of Oval's generality, the developers built emulations of several well-known CSCW applications — gIBIS, Coordinator, and Lotus Notes.

Lotus Notes combines features of both asynchronous collaborative applications, such as the two writing systems just described, and the network file systems, described earlier (Lotus, 1993).   Notes currently supports three basic functions: e-mail, text editing, and document management.   The e-mail and word processing functions are similar to those in other systems; it is the document management component that sets Notes apart from other applications.   Built to run on a network of PCs, Notes enables a group or organization to share a collection of documents that all members may access through the network and work on concurrently, within the constraints of security and a system of permissions.   For example, a member of a group can access a document, work on it, and return it to the storage system.   She can then send e-mail to a colleague telling him that it is his turn to work on it.   The second person can, in turn, access the newly revised document and do likewise.

Currently, Notes works best within local area networks, where all members may access the latest version of a document.   For groups distributed over a wider area, data must be copied from their primary storage locations and replicated on remote systems.   Users at remote locations, thus, work with replicated versions of documents, not the "originals."   Because changes to replicated data are updated among the various storage systems only occasionally, such as overnight, close coordination of work within widely distributed groups is difficult. Thus, Notes provides some of the function found in network file systems but with limited update and support for a limited number of data types.

Taken as a group, asynchronous collaboration tools and systems tend to be used with a group's more tangible forms of information. Because ftp transfers data from one file system to another, to the extent the files are a group resource, they represent some part of the artifact — either instrumental or target products. Private files obtained through ftp would represent either information outside the scope of the project, or, if the data contribute to a member's work on the project, they would become instrumental products. Network file systems have a similar relationship to the information flow model because they provide essentially the same access as ftp, but much more conveniently and with fewer restrictions (e.g., not requiring a login on remote systems). The same is true for collaborative applications, because they are used for direct work on the artifact.

E-mail, however, is used with all three major types, although not necessarily in equal proportion. Because most groups do not systematically save e-mail correspondence, messages tend to be ephemeral. But e-mail is also used to send copies of files between members and, thus, serves as a communication vehicle for both instrumental and target products. It can also affect intangible knowledge. To the extent that a message is substantive, it contributes to an individual's private store of intangible knowledge or the knowledge the receiver shares with (at least) the sender of the message. It can also contribute to knowledge shared by the group as a whole. For example, it is not uncommon in software development projects for individuals to make general announcements by e-mail, such as informing the group of a particular problem or that a new version of a module is available.

Thus, although asynchronous tools tend to be associated with tangible and ephemeral forms, some play important roles with regard to intangible knowledge as well.

Synchronous Tools

Synchronous tools enable collaboration by permitting two or more people who are separated geographically to interact with one another through a communication network. A large and diverse set of tools fall into this category. Some are concerned primarily with supporting verbal and visual interaction, such as two people engaged in a private conversation or a group having a meeting. Others enable several people to work together at the same time on the same tangible product, such a co-editing a document or drawing. In this section, three types of systems are discussed: audio/video conferencing, shared applications, and meeting support tools.

Most synchronous tools create working situations for distributed groups that approximate one or more "natural" activities that occur in groups whose members are all located at the same site. A few try to create new forms of interaction, based on new metaphors, such as two colleagues sitting opposite one another drawing on a vertical pane of glass that stands between them (Ishii, Kobayashi, & Grudin, 1992). However, they are the exception. Consequently, most tools are discussed in relation to the proximate forms of interaction they simulate.

Audio/Video   Conferencing

Audio/video conferencing tools allow members of a group to carry on informal conversations as well as more structured discussions at a distance. There are numerous analogs for this type of interaction: two people talking in an office, several colleagues having a discussion over lunch, a group holding a meeting in a conference room, or a programmer asking a question of a colleague sitting at a nearby workstation. Most such interactions have to do with intangible knowledge or with the social and administrative aspects of a project. Sometimes, they may also involve ephemeral products, such as a presentation that includes visual material. Occasionally, the group may combine verbal and visual interaction with work on the artifact, such as a group conducting a walkthrough of a new segment of code. But, by and large, these activities do not constitute direct and immediate work on the artifact; rather, they contribute indirectly — but essentially — to its development through the building of shared knowledge, through coordination, and by helping groups build and maintain a web of personal relationships and impressions.

The oldest, and by far the most widely used, tool to support verbal interaction at a distance is the telephone. It is so familiar and so taken for granted that most of us do not think of it as a tool for collaboration. But, we should not underestimate its importance for that purpose or as a standard against which to measure new alternatives.

To state the obvious, the telephone is most effective as a tool for two-way conversations. Many people use it reflexively for that purpose. Although they may be vaguely aware that talking on the phone and talking face-to-face involve different social and intellectual dynamics, most can do either with little or no conscious attention to the medium of interaction. Consequently, when two collaborators at a distance need to talk to one another, the telephone provides a level of service that is hard to beat.

We begin to notice the phone as medium when we use it in nonroutine ways. For example, calls that travel over great distances, such as between the United States and India or Australia, are often routed through a satellite. This can introduce a delay of nearly a second between the time one person says something and the other hears it. This latency significantly alters the natural rhythms of conversation. For example, the two speakers often talk over one another; when they realize this, each stops and waits for the other; hearing nothing, both respond, again talking over one another. Most people eventually adapt to the medium, often by exchanging rather formal statements and by signaling when each statement is complete. But, for many, it is not a natural or comfortable form of conversation.

A second nonroutine use of the phone system is for group discussions. Conference calls allow three or more participants to take part in a conversation. Each hears what the others say, and each may speak up at any time. But it is not the same as being in the same room. Because the group cannot see one another, they do not have access to visual cues, such as gestures or facial expressions, that signal when someone is about to say something, color what is said, or identify the speaker. Thus, although conference calls represent an important resource for distributed groups, the form of interaction this technology enables is different from same room interaction and leaves much to be desired.

To get around these problems, but also to provide additional capabilities, some groups have supplemented audio communication with real-time video channels. One of the earliest demonstrations of this was given by Doug Engelbart at the 1968 Spring Joint Computer Conference in San Francisco (Engelbart & English, 1968). During a plenary session, he and a colleague located some 30 miles away carried on a joint work session that included accessing a common store of information, editing a document together, and talking to and seeing one another through live, two-way audio/video channels. Much recent work in teleconferencing systems has been an attempt to make this mode of work more widely available.

In 1985, as an experiment in distributed organization, a group at Xerox's Palo Alto Research Center (PARC) was divided into two subgroups: one remained at PARC and the other was moved to Portland, Oregon (Abel, 1990). In addition to e-mail and high-speed data communications, members at the two sites were able to communicate with one another through "video walls" located in commons areas, such as lounges and coffee rooms. These set-ups consisted of large displays (although not really "wall" sized) and accompanying cameras and microphones, creating a form of "shared cross-site space" in which members could congregate and talk informally with one another. A similar system, called the VideoWindow, was developed about the same time by Bellcore (1989; Fish, Kraut, & Chalfonte, 1990).

Audio/video setups such as these seem to be most useful for informal discussions and for helping members of a group develop personal impressions of one another; however, they are less well suited for private, highly technical, or lengthy discussions. Consequently, during the latter stages of the PARC experiment, monitors and cameras were also installed in individual offices (Abel, 1990). Bellcore's CRUISER system was similar but used the computer to switch the video from one location to another, enabling users to browse one another's offices — located on several floors of a large building — as if they were walking, or "cruising," down a hallway (Fish, 1989; Fish, Kraut, Root, & Rice, 1992; Root, 1988). A more ambitious project was recently begun at Xerox's EuroPARC. They developed an extensive video environment, which they call *media space*, that includes each office at the Ravenscroft site as well as selected offices at Palo Alto (Gaver et al., 1992). What sets this system apart is the extensive set of applications they developed to supplement basic audio/video communication. They included a central calendar system that sets up video meetings and lists current or future meetings, CRUISER-like "hallway" browsing, and pages of "postage stamp" images of colleagues produced by slow-scan video updated every 5 minutes or so. This last application lets the group see at a glance who is around — within their 6,000 mile wide media space — and what they are doing.

To study the effects this technology on users, researchers at the University of Toronto have developed a system, called CAVECAT (Computer Audio Video Enhanced Collaboration At Toronto), which they use in both their own work and in more formal studies (Mantei et al., 1991; Sellen, 1992). The typical office set up of cameras mounted over participants' workstations with monitors located to the side does not provide realistic eye contact or a natural sense of spatial orientation. To counteract these effects, the Toronto group has developed an elegant device they call *hydra*. Each hydra unit, which resembles a carton of cigarettes mounted vertically on a swivel base, contains a speaker, a microphone, a color monitor, and a video camera located a fraction of an inch from the monitor. Because the camera and display are so close to one another, when a person looks at the display, he or she seems to be looking directly at the camera, creating strong eye contact with the viewer. During a meeting, several units can be placed on the desk in front of each participant to provide a sense of spatial orientation.

All of these are research or experimental systems, implemented using cable video technology and leased, high bandwidth communications lines. Consequently, they are expensive, one of a kind systems. Although still not cheap, equipment aimed at the corporate and consumer markets is beginning to appear. One such device is the codec. A rather compact piece of equipment, it uses digital compression to send video and audio signals over conventional dial-up telephone lines, although not yet at the NTSC standard 30 frames per second, making it possible to hold teleconference meetings in conventional conference rooms. AT&T has recently introduced three new forms of its Picturephone. Priced at approximately $1,000, these devices can be placed on individual desks or even in the home. Although this technology could make two-way video/audio communications much more widespread, we should be cautious in our expectations, given the unenthusiastic reception of the original Picturephone, introduced at the 1964 World's Fair.

To put this technology into perspective, I suspect that in the long run teleconferencing will prove useful in proportion to the computer applications that are available to control it, schedule it, and to provide services that go beyond simple "talking heads." If so, this would mean that the network that supports video/audio communication must be integrated with the group's computer network. This could be done in two ways. First, the video/audio signal could be compressed and transmitted as digital data over a conventional computer network, such as the Internet. Several computer applications are available that support telephone-like conversations over the Internet, but with considerable distortion.

Currently, large blocks of data are divided into pieces and sent over the networks in the form of packets. This is like transporting a large group of people from a hotel to a reception by putting small groups of them into individual cabs — they may all eventually get there, but not necessarily at regular intervals or in the order in which they left the hotel. Delays and irregularities in the arrival of video/audio information create distortions and latency. Given the intrusive effects of latency in distant phone conversations, as discussed earlier, this could severely limit or compromise the kind of audio/video communication that can be provided through current computer networks. One answer to this problem lies in advanced *isochronous* network architectures that provide pulse-like communication in which packets of information are delivered with guaranteed regularity. Although this technology is promising for widespread, high quality teleconferencing, using it for that purpose will require an extensive supporting infrastructure that rests atop the network's basic transport facilities. In effect, this would amount to developing a second telephone-like network within an upgraded version of the Internet or its successor.

An alternative approach is to develop more extensive connections between computer networks and the telephone network. Telephonic networks are circuit based, providing a virtual wire between endpoints with guaranteed bandwidth and latency characteristics. Thus, one can purchase one grade of circuit for two-way telephone conversations and another grade of circuit, with greater capacity, for full-motion video. At present, higher grade services are limited, costly, and awkward to set up and take down. But as fiber optic technology becomes widespread, network capacity will increase dramatically, and advanced telephone switches could provide dial-up high-capacity circuits adequate for teleconferencing, similar to current dial-up audio circuits.

If such an infrastructure were in place — and it seems probable that it will be relatively soon — it would be possible to build seamless collaboration environments that implement some of their services using conventional data networks, whereas other services, such as teleconferencing, would use telephonic networks. Each technology would be responsible for providing the services it does best. Although the current gap between data and telephonic networks seems large, most wide-area data networks are implemented using leased telephone lines — they simply attach specialized hardware that supports a different network protocol to high bandwidth telephone lines. Thus, dual-purpose networks could be built using the same underlying physical infrastructure.

As we look to the future, we should expect teleconferencing to become an increasingly available resource for collaborative groups. At first, it will be provided in separate systems; but, once established and once the connection is made between it and the computing infrastructure, a number of new applications should appear that will make video/audio communication useful in ways we cannot yet imagine. To make this happen, however, will require as much work in matters legal and political as in technical development.

Shared Applications

Shared applications permit two or more users whose workstations are connected by a communications link — such as a LAN, the Internet, or a telephone/modem connection — to work simultaneously and interactively with one another on the same data. Multiple users can provide input to an application program and see the results of that input on their respective displays. This mode of work is roughly analogous to several individuals sitting down at the same workstation, where they can all see the same display and where they can take turns using the keyboard and mouse. Normally, only one member of the group at a time is the active user — his or her workstation is the one that currently provides input to the system — while all of the others are passive users — they see the results of the active user's input on their respective displays. Most systems include floor control mechanisms that allow individuals to request to be the active user and for control to be passed among the participants. A few permit multiple active users in a free-for-all mode of work.

A typical task for which a group might use a shared application is co-authoring a document. The shared application could be a word processor or text editor. The group would set up one or more computer conferences in which the members participate from their individual workstations. During these sessions, all participants would be able to read the same segment of the co-authored document on their respective screens, and they could take turns writing or editing sections.

If during the session one member wants to make a comment to the others, he or she could type the comment into the document for the others to read. Some systems include chat windows, in addition to the shared application, that keep this type of informal interaction separate. More desirable would be to include audio and video conferencing, described in the preceding section, as an integral part of the shared application; but, currently, the two activities tend to be supported by separate and distinct systems. Thus, shared applications facilitate work on tangible products or, if the group does not save its work, ephemeral products, but they are awkward for group interactions concerned with intangible knowledge. As a result, the analogy of several people sitting down at the same workstation holds fairly well with regard to interaction with the application but breaks down with regard to the conversation such a group would normally carry on while they worked.

Although computer conferencing systems share many features, they differ from one another in subtle, but important, ways. To see these distinctions, let's look briefly at several example systems.

Aspects is a commercially available shared editing system that runs on Apple Macintosh computers using an Appletalk network or telephone lines with modems (Group Technologies, Inc., 1990). It includes three separate applications — word processing, drawing, and painting — that can be used separately or together in joint editing sessions. The three strongly resemble early versions of MacWrite, MacDraw, and MacPaint. Consequently, the price that must be paid for Aspects' collaboration support is accepting its very basic applications, as opposed to more sophisticated applications — such as MSWord, MacDraw Pro, and SuperPaint — members may be accustomed to using for individual work.

Aspects is an example of a *closed architecture system*. Other closed collaboration systems include ShrEdit, a collaborative text editor used primarily for studies of group behavior (McGuffin & Olson, 1992), and SEPIA, a system that includes planning, argumentation, as well as editing modes (Streitz et al., 1991). These systems provide fixed sets of tools, and the user is limited to those and only those tools. Should users want to work together using a different application, such as a project management tool, a scheduling program, or even a different text editor or drawing package, they could do so with these systems.

In contrast, *open architecture* conferencing systems provide general mechanisms that allow groups to work collaboratively with thousands of existing single-user applications, rather than with only the handful of specially tailored programs provided by closed systems. These systems currently operate within the UNIX and X Window environments and, as a result, are often referred to as *shared X* systems. The reasons for this are technical: the X Window architecture makes it convenient to intercept in a general way the input from mouse and keyboard directed toward an application as well as the output going from the application to the display. The conferencing system can then selectively redirect any given conference participant's input to the application, while blocking input from all the others; conversely, it can broadcast the output from the application to all participants' workstations. Thus, virtually any existing X/UNIX application can be used collaboratively, as well as new applications that will be developed in the future. Examples of shared X systems include Rendezvous (Patterson, Hill, & Rohall, 1990), MMConf (Crowley, Milazzo, Baker, Forsdick, & Tomlinson, 1990), Rapport (Ensor, Ahuja, Horn, & Lucco, 1988), and xtv (Abdel-Wahab, Guan, & Nievergelt, 1988).

Although shared X systems provide generality in the sense that a large number of single-user applications can be used collaboratively, they also impose restrictions. For example, all users see exactly the same display during conferenced sessions. Should one user want to scroll the window up to data not currently visible, he or she could not do so without first becoming the active user and then changing the information shown on all participants' workstations. An alternative would be a conferencing system that lets a group share a given application and data but lets each user control his or her display separately. Thus, a group could take turns editing a document, but while one person is editing, the others could look at different parts of it. Another possibility would be to allow multiple editors so long as two people do not try to edit the same paragraph. To support more flexible interactions such as these, Prasun Dewan developed a set of tools he called Suite for building advanced collaboration-aware applications (Dewan & Choudhary, 1991). Suite treats major collaboration design issues — such as the frequency users' screens are updated, the granularity of the data that are shared, and the restrictiveness of the coupling — as parameters that can easily be changed, permitting system designers to test different models of conferencing. Future systems could allow users themselves to control these parameters and thereby customize the system to suit their preferences.

To put conferencing systems into perspective, they represent a break with traditional systems in their goal of supporting groups working in close, continuous interaction on the artifact, as opposed to individuals working alone on separate data. For some activities — for example, a walkthrough of computer code, a review of a document, or a markup of legislation — working in a distributed manner using a conferencing system may actually be preferable to working conventionally in the same room. But this has not been demonstrated. Nor is it known with certainty for which activities computer conferencing is most appropriate, how those activities fit into the overall collaborative process, or the incremental benefits to be gained by adding human communication channels — audio alone and audio coupled with video. Thus, it is a technology that offers great promise but is still in its infancy.

Meeting   Support

Although computer conferencing systems are quite flexible and, thus, can be used in a variety of settings, they are oriented toward activities in which participants work collaboratively from their respective offices while doing some type of group editing task. A different set of tools has been developed to support situations in which participants gather in the same room at the same time to take part in a single, interrelated activity, such as a meeting.

A key part of most meeting support systems is a large, electronic display with facilities for various members to write, draw, or show computer and/or video data on the display. The early trend was toward setting up dedicated, specially constructed rooms. The display was provided by rear projection of video or computer data on screens mounted permanently in the walls. Participants controlled the display from individual workstations, often built into specially constructed conference tables or large, forum-style desks. Recently, a more flexible form of display has emerged that can be used in conventional meeting rooms or public spaces without physical alteration of the room.

Early meeting support systems tended to divide into two types. One group was oriented toward small meetings, typically involving six or eight participants, that would normally be held in a conference room. Most of these systems were built during the mid-1980s. Three examples are Capture Lab (Mantei, 1988), Xerox PARC's electronic meeting room built as part of their Colab project (Stefik et al., 1987), and a similar facility at MCC (Cook, Ellis, Graf, Rein, & Smith, 1987). The software provided by these systems ranged from simple editors, used off-line to enter text that was then cut and pasted to a display window, to relatively sophisticated hypertext systems that provided a spatial field on which the group could collectively build two-dimensional structures of ideas.

The expected mode of work was for individual members to take turns recording ideas on one of the workstations; the result would then be shown on the large display. Because participants were seated around a special conference table with embedded workstations, they could also see one another and talk both about ideas to be recorded on the group display as well as what was already there. Although developers attempted to minimize the effects of both hardware and software on behavior, all report significant changes in group dynamics. For example, if while one person was talking another began typing information on the screen, the group's attention often shifted to the display, including the speaker's who would sometimes forget what he or she was saying (Mantei, 1988). Thus, the display tended not only to focus discussion but to dominate it.

A second type of system is oriented toward large meetings involving several dozen participants. A well-known example is the group decision support system built at the University of Arizona (Nunamaker, Dennis, Valacich, Vogel, & George, 1991). Located in the School of Business, it supports tasks such as long-term strategic planning and policy formulation as opposed to small group discussions. The Arizona room, itself, has the feel of a corporate setting in its decor, lighting, and double rows of raised, forum-style desks that accommodate the 24 workstations used by participants. In addition to large display projection, the system also includes a number of applications that support group brainstorming, organizing, voting, and policy formation.

The typical mode of work is for participants to type messages into a shared database that can be read by all, send messages to authors of particular comments, and otherwise engage in multiple simultaneous on-line conversations. Thus, there is less emphasis on conventional discussion and more reliance on the system for exchanging views. However, at critical points, on-line activity can be interrupted, and the group can discuss a particular comment or data shown on one of the large displays. After that, the group may continue the current task, using the same program they were working with before the interruption, or they may move to a different task, using a different program. Implicit in the system, then, is a social/cognitive model of group intellectual behavior that encourages the group to move from broad, open-ended consideration of an issue, toward closure, to the drafting of some tangible product that records their collective thinking.

Recently, a third type of meeting support technology has appeared that is less intrusive and can be used in conventional conference rooms. Whereas the systems described previously assume that input is provided by a keyboard and, perhaps, a mouse, these systems are pen-based and assume that handwriting and hand-drawn figures are the major forms of input. Three examples include Liveboard, a custom hardware–software configuration developed at Xerox PARC (Elrod et al., 1992), Smart Technologies' System 2000 (SMART, 1993), and a research system developed at the University of Flinders (Mudge & Bergmann, 1993). However, just as *xerox* has become the generic name for copiers, *liveboard* is becoming the generic name for stylus-oriented systems based on the whiteboard metaphor. Consequently, I focus here on Xerox PARC's version of the concept.

Although Liveboard's developers look forward to flat, wall-mounted displays the size of actual whiteboards, currently, the unit is a large piece of furniture roughly the size of a console TV but twice as high. It provides an approximately 3 feet by 4 feet surface on which computer output can be displayed by a self-contained rear projection unit and on which users may "draw" using a special cordless optical stylus. The computer in the system is a conventional workstation — a UNIX workstation in research units, replaced by a PC in the commercial version. Although bulky, a Liveboard can still fit into the corner of many existing conference rooms and public places, such as lounges and coffee rooms, without physical alterations to the building.

The expected mode of work is similar to that for conventional groups meeting in conference rooms and using a conventional whiteboard. For example, the group might sit around a conference table discussing a problem or issue, while one or more participants stand in front of the display and write or draw on it with the optical pen, just as they would stand in front of a whiteboard and draw on it with a felt marker.   Currently, some half-dozen special purpose applications have been written for the Liveboard, including meeting tools, several whiteboard applications support for slideshow-like presentations, games, and a general interface to its internal workstation.   However, because the Liveboard screen can be used as a conventional workstation display, the group can also access documents or other products stored in computer files and work with them using conventional applications.   In the future, it should be possible to connect several Liveboards to one another through a computer network to support distributed meetings, although such meetings will, no doubt, require audio and, perhaps, video channels, as well.

The liveboard concept is a very interesting development. Designers seem to be on the right track to stress flexibility and hand, versus keyboard and mouse, input.   But, currently, it is a very expensive technology.   Much of its functionality could be provided at less cost by pen-based microcomputers or portable computers with stylus pads if they were all networked together and running a conferenced whiteboard application.   Hand-drawn input could selectively come from any of the computers, and output from the conferenced window could be displayed on each computer as well as on a large screen using inexpensive overhead projection equipment. The missing function would be direct drawing on the projected display, but it is hard to see this as a significant limitation if each participant could draw on his or her personal electronic tablet and have the result shown on the central display.   This configuration would also provide greater continuity between collective work that takes place in meetings and individual work that comes before and after.

Although only beginning to appear on the commercial market, meeting support tools could produce significant changes in the ways groups work by producing a change in the forms of information they work with in those situations.   In a conventional meeting, most of the interaction is verbal.   Usually there is no literal record of a meeting, and participants often disagree about what actually took place.   To help them iron out differences in their respective understandings of what they are talking about, groups often use ephemeral products — such as whiteboard drawings, foils, video, and so forth — that serve in the moment but are not kept permanently.   Even if a meeting focuses on some part of the artifact, as would occur in a design review or group editing session, the actual tangible product discussed is ephemeral — for example, printouts of code or documents maintained

in a computer system rather than the actual artifact.   After the
meeting, someone must update the permanent version, accordingly.
Because the literal product the group worked on is not preserved, this
can lead to disagreements about what was actually agreed to.   Thus,
most meetings are concerned with intangible or ephemeral forms of
information.

In a meeting that includes a meeting support system, products that
would normally be ephemeral, and hence lost, could become tangible
products that are stored permanently as part of the artifact.   For
example, whiteboard drawings that are normally erased could be
stored in computer files if they were produced on a Liveboard rather
than a whiteboard.   During a meeting, the group could also address the
artifact directly, rather than ephemeral products derived from it.
Thus, groups using meeting support systems should show a significant
shift toward more tangible forms of information and more focused
discussion.

It would seem obvious that a change that makes discussion more
tangible would be beneficial, but this may not always be true.   For
example, during a series of brainstorming sessions, such a shift could
lead to premature closure around an early position that is retained in
tangible form and repeatedly brought back into the discussion,
whereas a conventional, less tangible discussion might have led to
consideration of a wider range of options.   Additional research is
needed to understand the many subtle ways these tools will affect
collaborative behavior and to help groups make effective use of them.

A Comprehensive System

In the two preceding sections, I described tools that support the
independent, asynchronous work of groups as well as tools that
support their real-time, synchronous activities.   Groups, however, do
not make clean distinctions between these two modes of work.   For
example, an individual working alone may call a colleague into his or
her office to ask that person's opinion about a paragraph he or she is
working on.   Or, a question addressed to one individual across a
computer lab can lead to an informal discussion joined by several
other people working nearby that, in turn, leads to a search for a

particular piece of code that the group then discusses. If each activity is supported by a different system, shifting from one to another may be awkward, and moving data from one context to another may be difficult or impossible. Instead, groups need systems that span the spectrum of synchronous and asynchronous activities and permit quick, easy transitions from one mode of work to another. More specifically, they need comprehensive systems that combine most of the individual tools described previously and can accommodate new ones not yet developed.

Over the past several years, the UNC Collaboratory Project has built a system that is trying to meet this need. Called the Artifact-Based Collaboration (ABC) system, it supports development of the artifact by groups engaged in both synchronous and asynchronous work (Smith & Smith, 1991). It also includes tools for creating and using ephemeral products and for developing shared intangible knowledge. Thus, it supports the three basic types of information included in the information flow model, although not necessarily all situations in which each occurs.

In Part II, it will be easier to discuss collective intelligence as a form of computer-based behavior if we can assume that groups have access to a particular set of tools. Consequently, I use ABC as the example system that provides that basic set. However, ABC is not the only system that supports multiple activities; others include Doug Engelbart's Augment (Engelbart et al., 1973), whose legacy underlies virtually all work in this field, Mermaid (Watabe, Sakata, Maeno, Fukuoka, & Maebara, 1990), SEPIA (Streitz et al., 1991), and HB1 (Schnase, Leggett, & Hicks, 1991).

Overview of ABC

Figure 3.1 provides an overview of ABC. The system is designed to be used by distributed groups whose members may be located a considerable distances from one another, clustered at a single site, or combinations of clusters and scattered individuals. Thus, a key assumption is that members do much of their work through workstations connected to a high-speed network. This network connects each workstation with a *hypermedia data storage* facility as well as with other workstations. Users work with *browsers* that

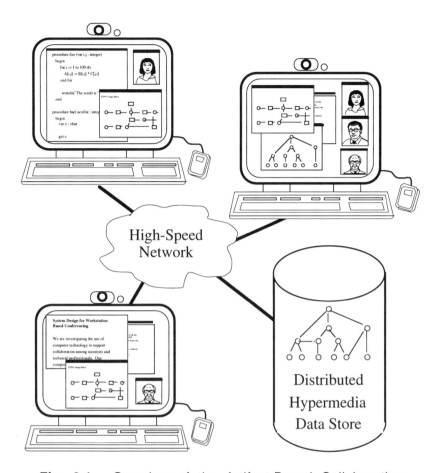

Fig. 3.1. Overview of the Artifact-Based Collaboration (ABC) System. Three workstations are connected to a hypermedia storage system and to one another by a high-speed network. The same conferenced browser can be seen on each workstation, with supporting video windows. Other nonconferenced browsers and applications can also be seen on each workstation.

enable them to see and manipulate structural data and with *applications* that enable them to work with traditional "content" data, such as text and diagrams. A *conferencing* component allows several members to share any browser or application so that all can see the same information and each, in turn, can edit or manipulate the data. The design also includes *audio and video* communication so that members can talk to and see one another, both to supplement computer conferences and for conventional conversations. Thus, ABC integrates support for three major forms of collaborative work — individual work on the artifact, collective work on the artifact, and conversations/discussion — thereby supporting both synchronous and asynchronous activities.

To understand ABC more clearly, one must understand five key concepts or system features: virtual screen, hypermedia data store, browsers and applications, conferencing, and audio/video communication. Each of these is discussed later.

Virtual   Screen

ABC runs within the X Window System under the UNIX operating system. However, the data users work with is stored in the hypermedia data storage system, described later, rather than the UNIX file system. This requires users of ABC to have a different mental model of their data and to use different tools to work with it. In addition, the system provides several kinds of generic function, such as conferencing and hyperlinking, that apply to any program or application running within it. Consequently, it is important for users to be aware of whether they are working within UNIX or within ABC at any given moment.

To do this, ABC provides a facility called a virtual screen (Jeffay, Lin, Menges, Smith, & Smith, 1992). It is a window with an identifying label. However, it can include other windows, and it can be expanded to cover the entire screen. All ABC programs run within this virtual screen, and any program that runs within it references the hypermedia data store, rather than the UNIX file system. Access to ABC generic functions is provided by an additional menu bar attached to the top of each program window within the ABC virtual screen.

Thus, ABC can be viewed as an environment within the larger UNIX context, or it can be expanded to appear to be the entire system.

Hypermedia   Data   Store

ABC encourages a group to think of the entire collection of information it builds and works with as a single, integrated structure, rather than as a set of separate and independent files (Shackelford,

Smith, & Smith, 1993). This structure constitutes the group's artifact. For a software development project, the artifact might include early concept papers, requirements, specifications, the architecture, source code, maintenance manuals, user documentation, and perhaps conference papers and journal articles that describe the system. It could also include the private or personal data of the individual members of the team.

In terms of the ABC data model, the artifact consists of a collection of separate *graphs* that are composed to form a single large, interrelated structure. Currently, ABC supports three types of graphs: *trees, networks,* and *lists.* Each separate graph corresponds to some presumably logical entity, such as a short document or a section or chapter of a larger document. Graphs, in turn, consist of sets of *nodes* and *links.* Nodes normally represent concepts, whereas links represent relationships between concepts.

Some graph types are better suited for particular tasks than others. For example, one could use a tree to represent the hierarchical structure of a document, similar to an outline. The title would appear as the label of the node at the top, nodes for major sections under that, subsection nodes below each section, and so forth. However, because tree graphs permit only links from a "parent" node to its "child" nodes, they do not allow links between "sibling" nodes or links that would otherwise cross the hierarchy. If more flexibility is needed, for example, to record ideas and relationships generated during a brainstorming session, one can use a network graph, which permits links between any pair of nodes.

An important concept in ABC is node *content.* Although nodes can be used to represent concepts directly, they can also contain two types of information. First, nodes can contain a block of data similar to a conventional file. For a short document represented as a tree, the actual text, diagrams, or other forms of information that constitute its substantive content can be stored as the contents of the "leaf" nodes at the bottom of a tree. A printed version of the document can, then, be derived by having the system go around the tree and gather up all of the pieces stored as node contents and send them to a printer. Second, nodes can contain graphs. That is, separate, disjoint graphs can be stored as the contents of nodes, just like the blocks of data described previously. For example, a group could collectively develop the overall structure of a document as a two-level overview tree that identifies the document as a whole and the chapters to be included in

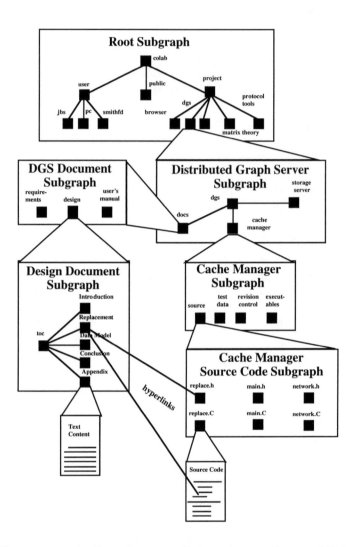

Fig. 3.2. Artifact for a collaborative project. Higher level nodes contain graphs, whereas nodes in lower level graphs contain file data, in this case, text. Hyperlinks connect a node in one graph with nodes in another, one of which is anchored within the node's content.

it. They could then work individually or in teams on the detailed plans for each chapter as separate graphs that are contained within the chapter nodes. Indeed, this pattern of work was seen in several of the scenarios described in chapter 2. By successively deeper nesting of graphs, groups can build structures that are very large, but ones that can be viewed at different levels of detail, making them easier to understand and easier to work with. Figure 3.2 illustrates how a group might organize all of its information as a single artifact.

Finally, the ABC data model includes a second type of link, called a *hyperlink*. The links described to this point are *structural* links. They are constrained by the rules that apply to a given graph type. For example, a node in a tree graph can have (at most) one incoming link, but any number of outgoing links. This constraint, although helpful in some contexts, precludes relationships that are also desirable, such as cross references, that would violate the basic tree structure. Hyperlinks provide this flexibility without violating the integrity of the graph type.

Hyperlinks can define two types of relationships that cannot be defined using structural links. First, they can join nodes (or points within the contents of nodes) to one another within a graph that would violate the type of the graph if the relationship were defined using a structural link. Second, hyperlinks can join nodes in one graph to nodes in another graph. Because a structural link must exist within some particular graph, they cannot be used for this purpose. This second use of hyperlinks is especially important for collaborative work. As discussed in chapter 2, maintaining the internal consistency of the artifact becomes a major problem for large projects. Hyperlinks provide a mechanism through which groups can identify dependencies within the artifact so that when a change is made in one place — for example, a specification document — one can follow hyperlinks to other places within the artifact — for example, the source code and user documentation — that may be affected by the change and, thus, require update or verification. Figure 3.2 shows both types of hyperlinks.

Browsers and Applications

Members of a collaborative group work with ABC through two types of tools: browsers and applications. Browsers enable users to view and work with the structure of the artifact. Through them, users create new graphs as the contents of nodes; add, delete, and move the nodes within a graph; create and edit links between nodes; add or change identifying labels on nodes; or simply view an existing graph. Currently, ABC includes browsers for trees, networks, and lists, corresponding to the three types of graphs supported by the data model. Other browsers could be developed, such as a decomposition diagram editor, whose underlying data model is a graph but whose appearance differs from conventional graph representations.

Applications are used to work with blocks of nongraph data. Examples include text editors, drawing programs, spreadsheets, CAD/CAM tools, and so forth. ABC provides an open architecture with respect to applications, so long as the application operates within the X/UNIX environment. Thus, a group can use ABC browsers for working with the structure of the artifact, but the tools it is accustomed to using for working with the individual files/blocks of data contained within the graph structure.

Conferencing

ABC supports shared X conferencing as one of the generic functions it adds to any browser or application running within it. Conferencing is accessed through the second title bar ABC adds to a virtual screen and to all browsers and applications that appear within it (Jeffay, Lin, Menges, Smith, & Smith, 1992). This allows members of a distributed group to work together at the same time on a document, drawing, or other form of information by sharing any browser and application running in the ABC environment. They all see the same display, and they can take turns providing input to the conferenced program.

One unusual feature of ABC conferencing is that it permits multiple browsers/applications to be shared within a single conference. This can be done by conferencing a virtual screen and all browsers and applications running within it, as opposed to a single program. Thus, for example, if a question comes up during a conference that cannot be answered by reference to currently shared data, a member can start a browser or application on other data and then bring the second program into the conference.

By integrating conferencing into the collaborative environment, ABC enables members of a group to shift easily and smoothly between individual, asynchronous work to synchronous, collective work. And, of course, there is no issue of moving data from one context to the other, because the group works with portions of the artifact in both situations.

Audio and Video

Although computer conferencing can help distributed groups work together on the artifact in ways that would not be possible otherwise, members also need to talk with one another, hold meetings, and carry on both formal and informal discussions. No technology can replace face-to-face encounters. But it may be able to supplement those encounters, particularly for groups that are widely distributed and for whom frequent face-to-face interaction is not possible.

ABC includes audio and video communication within its basic design. Currently, this function is implemented outside the ABC environment proper, using the telephone system for audio and cable television technology for video. In the next stage of our work, we plan to fully integrate audio and video communication into ABC. To do this, we will build on work done by my colleague, Kevin Jeffay (Jeffay, Stone, & Smith, 1992, 1994). Using specialized scheduling algorithms developed by his multimedia research group and conventional data compression hardware, Jeffay's system can deliver full-motion video and high-quality audio over conventional packet-switched data networks. We plan to incorporate this type of digital audio and video as a generic function, analogous to conferencing, that can be accessed from within ABC and controlled by ABC applications. Thus, users will be able to schedule and setup audio/video conversations analogous to starting a computer conference, enabling them to make smooth transitions from individual work, to conversations with colleagues, to video supplemented computer conferences.

Thus, ABC includes a broad range of functions that are normally found in separate collaboration tools. These functions address all of the types of information included in the information flow model as well as key transformations. ABC also supports both synchronous and asynchronous activities, and the system permits easy transitions from one mode of work to the other.

For asynchronous work, ABC allows distributed groups to work within a single, integrated structure of information — the artifact — with the sense that data is omnipresent, rather than a scattered collection of files that must be moved from one location to another for access. It includes basic facilities with which to note dependencies and other relationships within the artifact that must be maintained if the structure is to be coherent and internally consistent. And its data model provides a natural means of partitioning the artifact for ease of understanding, for supporting multiple concurrent users, and for scalability.

For synchronous work, ABC includes computer conferencing as a generic function that can be applied to any browser or application. When fully implemented, ABC will also provide audio and video communication over conventional data networks. Currently, ABC does not include explicit meeting room support; however, whiteboard applications and other similar meeting tools could be run as ABC applications and used collectively during meetings. In the future, we plan to explore using ABC with new input/output technologies, such as stylus pads, pen-based computers, and liveboards.

Groups have worked in the past and, no doubt, will in the future using conventional computer and communications systems as well as more basic tools, such as pen and paper or typewriters. However, as systems are developed explicitly to support collaboration, it seems safe to assume that they will play a larger and larger role in collective work. For the remainder of this discussion, I assume that collaborative groups are using ABC or an equivalent system so that I can assume they have access to a specific set of capabilities, thereby making it easier to define collective intelligence as a form of *computer-mediated* behavior. Later, one could go back and factor out ABC-specific features to generalize the concept.

Issues for Research

There are a number of issues and questions concerning collaboration systems that need further research. Here, I briefly discuss several that seem particularly important.

- *What is the relationship between the process of collaboration and systems that support that process?*

We need to look closely at the relationship between the needs of users and the services provided by systems. Which collaborative activities are supported by existing systems? Do those tools help groups work better or more efficiently? Are there other collaborative activities for which no supporting tools exist? As better models of the collaborative process are developed, we should continuously map the activities and the information transforming processes identified in those models against supporting tools, both for purposes of evaluation and for identifying gaps where new tools might be built.

- *What should the user interface for collaboration tools be like?*

Should the user interface for a collaboration system be different from the interface for a system intended for individual use? Currently, many systems base their user interface on the metaphor of the desktop. Because a desk is normally used by an individual, is the desktop an appropriate metaphor for collaborative systems, or is there some other metaphor that would be better? Do we really need such a metaphor? The relationship between everyday objects and computer functions quickly breaks down; consequently, metaphors are most helpful for novice users and can become counterproductive for advanced users. We may be better off identifying a common set of user functions needed by most collaborative applications and then presenting them to the user though some consistent, abstract, but nonmetaphoric image, such as a graph.

- *Which facilities can help groups organize and access information in wide area storage systems?*

Which form of wide area storage system is most appropriate for distributed collaborative groups: wide area file systems, distributed database management systems, network accessible document management systems, or closed architecture hypermedia systems? Do different groups need different storage systems, or will one type dominate in the long run? If wide area file systems become predominate, they will enable users to access large, widely distributed, and, most likely, unstructured collections of files; can and should these collections be given more coherent form? If so, is a hypermedia graph model appropriate, or is some other structure better? Should multiple groups be able to construct multiple structures over the same files? What new problems of consistency and security would this raise? Is replication an essential requirement for performance and, if so, how frequently should replicated data be updated? What sort of browsing and search functions are needed? Does the user need to visualize the information structures he or she works with and, if so, what tools can help?

- *What kinds of interaction are needed among computer applications and processes?*

What should the system and communications infrastructures provide to support shared editing and other forms of real-time application-based interaction? Should support be provided at the level of the window manager or at some deeper level of the system? Are existing protocols, such as X Window, adequate, or is a new protocol needed that assumes collaborative interaction as a basic requirement? What is the proper level of granularity of interaction? Should granularity be a developer or end-user option? How frequently should data be updated? Should all users see the same view (i.e., WYSIWIS: what-you-see-is-what-I-see), or do users need independent views? Are there other conceptual models for real-time application-based interaction in addition to shared editing?

- *How can human and data communications best be combined?*

How can human verbal and visual interaction best be supported among distributed groups? Should support for voice and video be treated as separate services, or should they be integrated into the workstation so that they can be used in conjunction with traditional data applications? Should the phone company or should data networks such as the Internet provide these services? That is, should support for voice and video be developed within data communications protocols and architectures, or should the emphasis be placed on developing better interfaces between computer and telephonic systems? Because most data networks are implemented using leased telephone lines and, conversely, because most telephone systems are digitally based, could dual purpose networks be achieved through standards? If it turns out that human interaction is best supported through data networks, what changes in computer and communications architectures will be needed to address fundamental problems of latency, synchronization, and reliability?

- *What would future systems infrastructure look like if support for collaboration and cooperative work became a basic requirement?*

Virtually all existing computer systems were designed to support individual users working alone. This is obviously true of personal computers and individual workstations. But it is no less true of large, mainframe systems that provide their numerous users with the illusion that each is the sole user of the system (e.g., CMS' concept of the virtual system). Early collaboration systems were treated as applications, supported by this infrastructure. More recently, a few have taken a more general approach by adding architectural layers to the infrastructure, such as a graph abstraction layer on top of distributed file systems or shared application support on top of window management protocols, with the result that they can provide collaboration support for open-ended sets of applications. But, these more general systems are often forced to compromise or to solve problems in ad hoc ways, because they rely on existing operating, file, window manager, and communications systems, none of which were designed with the needs of highly interactive, widely distributed collaborative groups in mind.

What if we began with a blank sheet of paper? What fundamental changes would result in system and communications architectures if we assumed that support for collaboration and communication among users was a *fundamental* requirement, on the same level as providing permanent storage, allocating resources, managing main memory, providing interprocess communication, and so forth? What would *that* infrastructure look like? Once conceived, could it be implemented on top of existing systems and communications infrastructures, or would it require building from the ground up? If it turns out that the second alternative is needed, could this be done given the enormous legacy of existing hardware and software?

Chapter 4

Cognitive Models and Architectures

In chapter 2, we saw that different collaborative groups can be characterized according to the different types of information they use and produce over the course of a project and the flow of information they generate in transforming one type into another. In chapter 3, several kinds of computer and communication tools were discussed that support these information processing activities. In this chapter, I look at models and architectures that describe human cognition as a form of information processing system. These three perspectives, then, provide the basic materials needed to build a concept of collective intelligence as a form of computer-mediated behavior in which human beings supply the mental processes used to build large, complex structures of ideas.

The chapter is divided into three sections. In the first, I review general cognitive models and architectures. Next, I look at specialized models or frameworks that apply these ideas to particular tasks or situations encountered by collaborative groups. In the third section, I look briefly at an objection raised by Allen Newell to the idea of group intelligence. The goal of the first two sections is to identify a basic set of components, broader than any single theory, that have been used to describe human mental function. In Part II, I identify constructs within computer-based collaborative groups that are recognizable as extrapolations of these components as a starting point for building a concept of collective intelligence. I also try to answer Newell's objection.

General IPS Models and Architectures

For the past 20 years, cognitive science has been dominated by the view that human cognition is fundamentally concerned with the processing of information. In this section, I review Newell's and

Simon's original Information Processing System (IPS) model that established this perspective. After that, I look at two more recent models that sum-up much of the intervening work. These include Newell's IPS-based cognitive architecture and an alternative architecture developed by John Anderson.

IPS

The original Newell and Simon model explains how human beings carry out complex problem-solving tasks (Newell & Simon, 1972). Their goal was not to develop an abstract cognitive model as an end in itself but, rather, to develop computer programs that could simulate human intelligence. They based their model on think-aloud protocols of human subjects solving three types of problems — cryptarithmetic, theorem proving, and chess — but they also incorporated a number of other results from the literature into the model.

A key assumption in their work was that cognition is inherently concerned with processing information. Hence, their model emphasizes the representation, processing, and storage of information and has come to be called the Information Processing System perspective. A high-level view of this model is shown in Fig. 4.1. Its major components include *Receptor* and *Effector* functions that interact with the external environment, a *Memory*, and a *Processor* that operates on information flowing among these components.

Because the entire system is oriented toward the flow and control of information, a key concept is *symbol*, the primitive unit of information on which and by which the Information Processing System works. Although the model suggests that effector and receptor functions map between external phenomena and symbols internal to the system, these components are not developed in any detail in the model. Rather, information is directly inserted into the system or obtained from it by the researcher. Thus, the model is most concerned with the processing of information after it already exists in symbolic form within the system.

Although symbols can be stored individually in memory, they are normally organized into *symbol structures*. These structures may be *designative*, in which case they denote semantic information, or they may be *programs*, in which case they represent operations or *methods* that can be applied to symbol structures to derive information about them — for example, compare two structures for equivalence — or to change them — for example, add a new symbol to the structure. Designative information is stored in the memory as associative structures of symbols organized into hierarchical chunks. Methods, on the other hand, are stored as production rules.

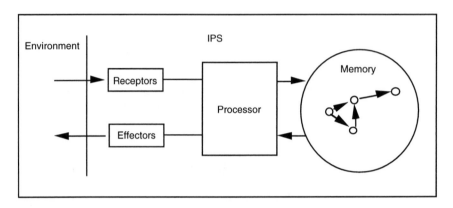

Fig. 4.1. Newell and Simon's Information Processing System (IPS) model of cognition. (Adapted by permission from Newell & Simon, 1972.)

The processor provides the context in which elementary cognitive processes, represented as production rules, operate on the semantic contents of memory, represented as a graph structure. Not shown in the figure is a short-term memory (STM) component that is part of the processor. It provides a cache that is loaded from memory as well as from perceptual mechanisms that are part of the receptor. Processes operate on the symbols/symbol structures within it; subsequently, its contents can be encoded and stored in the memory.

The IPS model becomes animated through the process of solving problems. A problem is presented to the system with respect to an external *task environment*. However, the system does not operate on the problem within that external context but, rather, within an interior *problem space*. The problem space contains a goal — the solution to the problem — a representation of the problem, a representation of relevant aspects of the task environment, and a data structure consistent with these representations. Methods are applied to the initial representation of the problem, producing incremental changes in the data structure. The process is repeated until a path is constructed that joins the initial problem to the data state that represents the goal. However, if along the way the processor reaches an impasse and cannot complete the construction, it generates one or more subgoals in an attempt to resolve the impasse. Each new subgoal, in turn, generates a new problem space. Frequently, the new problem space requires translating the problem or subproblem into an alternative representation appropriate for that context, and it is often this change in representation that enables a solution to the problem. Once the subgoal is achieved, processing returns to the higher level problem space in which the subgoal was generated and continues from there. This hierarchical nesting is both recursive and iterative and, in theory, unlimited in the number of levels of descent that can be generated.

From this brief description, we can identify several basic components found in the IPS model. These include the memory, the processor, and its short-term memory component. Effector and receptor functions are also included, but they are not developed in any detail in the model. At a more abstract level, the problem-solving system also includes goals, problem spaces, and their internal representations and data structures. Although these latter constructs play key roles in human problem-solving activities, Newell and Simon regarded them as task-dependent and, thus, informational objects learned by the individual problem-solver rather than as basic components of the IPS model.

Soar

In the 20 years since the publication of the original IPS model, Newell's views of human cognition obviously changed. But what is most striking is not the differences in those views but their continuity. The changes represent extrapolations and refinements, rather than completely different formulations. Newell's more recent goal was to develop a comprehensive cognitive architecture that could provide a framework in which to develop unified theories of cognition — systems that incorporate the sum of what is known about human cognition (Newell, 1990). The vehicle for supporting such theories Newell called *Soar*.

Soar is several things at once. First, Newell emphasized that Soar is an architecture, rather than a model. By that he meant that Soar includes as basic constituents those mental functions and constructs that remain fixed under different tasks and conditions. For example, the mechanism that performs memory accesses is regarded as invariant and, thus, architectural, whereas the contents of a particular memory access is variable and, thus, data. This separation between general and specific makes it possible to define different cognitive models within the general Soar framework. Second, Soar is a computer programming language in which to write Artificial Intelligence (AI) programs. It is based on the OPS-5 expert system language but includes several enhancements, including a problem space construct and chunking mechanism that are basic parts of the Soar language. Third, Soar also refers to individual computer programs, written in the Soar language, that simulate specific cognitive tasks and processes or implement specific cognitive models. Soar programs have been written that duplicate the original problem-solving tasks described in Newell and Simon (1972), perform syllogistic reasoning, verify elementary sentences, acquire skills through experience, and simulate an impressive list of other cognitive functions.

Although Soar bears a strong resemblance to the IPS model, it is also different in several important ways. In the IPS model memory and processor were separate components, and the processor contained within it a small working cache, called the short-term memory. In the Soar architecture, shown in Fig. 4.2, memory is divided into two components — long-term memory and working memory. Furthermore, the processor has disappeared as a separate component and is replaced by a set of basic cognitive processes, a more abstract concept. These processes are stored in long-term memory but operate within the new working memory component. Another major distinction is that the associative network structure of long-term memory in the 1972 model has been replaced in the Soar architecture by additional production rules. Thus, both declarative knowledge as well as cognitive processes are represented as productions.

A Soar system becomes animated through a process of *searching*, similar to that in the IPS model. Operators or methods are applied to a data structure in order to achieve a goal state within a problem space. Thus, all processing takes place within the context of the particular problem space that is currently loaded into the working memory component of the system. The problem space construct is similar to that for the IPS, but in Soar it is regarded as a fundamental part of the architecture that applies to all human beings rather than as an ad hoc information structure learned by an individual. Thus, all processing is cast within the general paradigm of constructing a path from the initial problem state to the goal state and of resolving the impasses that arise along the way. However, a major difference is the way Soar learns from resolving impasses. Once an impasse is resolved, Soar engages a process called *chunking*. A new production rule is created that records, as the conditional part of the rule, the state of the problem space at the time the impasse was encountered and, as the action portion of the rule, the path to solution. Chunking is Soar's primary learning mechanism and is a basic part of the Soar architecture, rather than an ad hoc procedure.

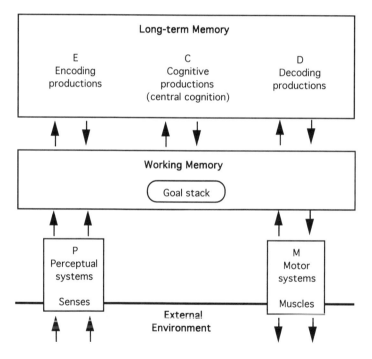

Fig. 4.2.   Newell's Soar cognitive architecture. (Adapted by permission from Newell, 1990.)

Thus, Soar has retained the IPS's long-term memory component and mechanisms for interacting with the external world, although these latter components are still undeveloped as they were in IPS.

However, important changes have been made in the structure of long-term memory and in the disappearance of the processor component. Both semantic and procedural knowledge are stored as production rules in Soar. The processor has been replaced by a working memory component that provides the context in which processing takes place. Although particular problem spaces are task specific and, hence, informational, the underlying form common to all problem spaces is invariant. Consequently, both problem spaces and the primitive functions that operate on them — to select a particular problem space, to select a data state within it, and to select and apply operators to that state — are considered to be basic parts of the architecture. Finally, the impasse resolution and chunking functions provide the principle mechanisms for learning and give Soar much of its generality and flexibility. They, too, are regarded as part of the architecture.

Act*

A second comprehensive architecture has been developed by Newell's long-time CMU colleague, John Anderson (Anderson, 1983, 1990). Newell called Anderson's Act* the first unified theory of cognition (Newell, 1990). It grew out of Anderson's earlier work on the semantic structure of human memory, modeled as an associative network (Anderson & Bower, 1973). Since then, Anderson has added a process component, in the form of production rules, that acts on the contents of memory and a skills acquisition feature that allows rules to perform actions on other rules. Like Soar, Act* is implemented as a computer system. Act* programs can simulate a wide spectrum of mental behaviors ranging from basic cognitive processes, such as fact retrieval and stimulus-response behaviors, to higher level functions and problem-solving tasks, such as language acquisition and constructing geometric proofs.

The Act* architecture, shown in Fig. 4.3, is similar to Newell's Soar system in its interactions with the external environment and in its distinction between long-term memory and working memory. But it differs from Newell's system in several important respects. Among these differences are its division of long-term memory into two separate storage systems, its incorporation of activation as a fundamental part of the architecture, and its inclusion of multiple data types for individual node contents within the declarative memory.

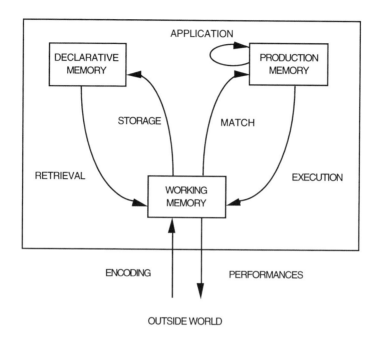

Fig. 4.3. Anderson's Act* cognitive architecture. (Adapted by permission from Anderson, 1983.)

Anderson divided long-term memory into two separate stores, which he called *production memory* and *declarative memory.* Declarative memory is the store for facts and is structured as a semantic network. Nodes in the graph store information as short sequences, images, and propositions; more complex structures are composed of hierarchies of these basic lower level types. Thus, there are different storage subsystems in the declarative memory for different types of information. Nodes are joined to one another through associative relations and each has a level of activation. Production memory is the store for processes, represented as production rules. These rules operate on the contents of declarative memory, but they also operate on the contents of the production memory itself to produce new rules, modify existing rules, and so on.

Working memory is the context in which processes are applied to the contents of the declarative and production memories. Thus, it functions as a form of problem space, but unlike Newell's systems that include multiple problem spaces, it is the sole problem space in Anderson's architecture.   The figure is misleading with respect to working memory in one important respect. Working memory is not a separate component in Anderson's system but, rather, that portion of the other two memory systems that is currently activated. A process of spreading activation determines which parts of the two memory systems are the working memory at any given moment. As processing takes place and, thus, activation levels and conditions for rule selection change, so working memory changes. By analogy, think of looking down on a darkened surface — the long-term memory systems — and sweeping a flashlight across that surface, selectively illuminating different parts — activating or condition selecting different parts of the memory stores. The currently lighted/activated/condition-selected parts constitute the working memory at that moment. This concept of working memory stands in sharp contrast with the view that it is a separate component into which contents are copied from long-term memory and vice versa.

Since the original description of the Act* architecture in 1983, Anderson extended the capability of the system so that it can formulate more sophisticated strategies and adapt its behavior to changes in the environment (Anderson, 1990). This is done through a three-step process in which the goals of the system are first determined, then a formal model of the environment is developed, and, third, the computational limits of the system are identified. From these three sources of information, an *optimal behavior function* is defined for the system in its current context relative to current goals that is then applied to the relevant problem or task. Because the process may produce errors, it can be iterated until the system produces behaviors that are consistent with empirical results. Thus, the additional level adds a form of *rationality* to the Act* architecture.

In summary, Anderson's Act* system grew out of his earlier work on human memory. It is similar to Newell's Soar system in several respects. It represents cognitive processes as production rules. It employs a type of problem space construct as the context for higher level functions. And it controls its behavior through a hierarchical goal-subgoal structure. However, Act* differs from Soar in several ways. It includes two separate memory systems — a production memory for processes and a declarative memory for semantic information. The declarative memory is organized as a network of nodes whose contents are represented in terms of a small set of basic data types. Activation plays an important role in Act*. Nodes and productions have associated activation values and a spreading activation process figures prominently in the architecture. Thus, its working memory is not a separate component but the currently activated portion of its memory systems. Although Anderson's recent work in rational analysis may lead to a more sophisticated concept of strategy, that remains a weakness in the Act* architecture as it is for other IPS systems. Act* is similarly limited with respect to perception and motor actions.

From these three IPS models/architectures, we can sketch a composite view of cognitive systems and their components. If we wish to regard collective intelligence as an IPS system, it, too, is likely to have components similar to these or components that are at least recognizable as extrapolations of them.

The system is likely to include a long-term memory component, probably represented as some form of graph structure and/or as a set of production rules. If the long-term memory is modeled as a graph structure, nodes within it may contain relatively small amounts of typed data or (hierarchical) structures of lower level nodes. The system should include a working memory, either as a separate component or as that (small) part of long-term memory that is currently activated. And it should include either a separate processor component or, more likely, a set of processes that operate on the activated contents of working memory.

Higher level tasks — for example, problem-solving — are likely to be performed within the context of one or more problem spaces or some similar architectural construct that includes a goal, a scheme or data type in which to represent problems, and a data structure consistent with that representation. To chart an overall approach to a task and to respond to problems or altered conditions within particular contexts, the system should include sophisticated strategies and tactics. Although solving a problem the first time may involve considerable trial and error, with experience, the system should learn to recognize problems and to retrieve solutions worked out earlier for similar problems.

Finally, a complete cognitive system would also include receptor and effector functions that interact with the outside environment, although these functions have largely been ignored in IPS models and architectures.

Specialized IPS Models

In the preceding discussion, I reviewed three general cognitive models and architectures. These systems have been most successful at modeling or simulating isolated forms of human behavior, such as memory access or stimulus-response behaviors; at solving well-defined problems, such as cryptarithmetic and chess; and at solving problems in which weak principles of strategy, such as hill climbing and means–ends analysis, are sufficient. For problems of this type, one can define a set of rules that can be used in a systematic way to carry out the task, and one can tell unambiguously whether or not a solution has been reached. Consequently, these models have routinely been expressed as computer simulation programs.

They have been less successful at modeling or simulating coherent behaviors that extend over long periods of time; at solving poorly defined problems that require judgment and/or qualitative evaluation; and at performing tasks that require complex strategies or interactions with other individuals. Knowledge-construction tasks — such as writing documents and computer programs, designing a building, or planning a sales campaign — are typical of this type of activity. The rules for performing these tasks are usually general — more like rules of thumb rather than precise procedures that can be systematically applied — and no rule exists to tell exactly when the task is complete. They also require strategies that extend over hours, days, or even longer periods of time. Currently, the state of the art does not support models for extensive conceptual construction tasks that have the rigor and consistency of those for well-defined problems. Imagine, if you will, what would be required to develop a simulation system that could write a journal article, a proposal, or even a letter to a friend.

Nevertheless, significant progress has been made over the past decade in understanding complex, real-world forms of cognition. Researchers have developed specialized models and frameworks that apply to specific tasks, such as expository writing, or to specific situations, such as carrying out a particular task using a particular computer system. They have incorporated concepts from the more general systems and architectures discussed previously, demonstrating the applicability of those concepts to ill-defined problems. But they have also been forced to extend those concepts, change their points of view, or accept limitations in their research agendas.

In this section, I extend the discussion of basic components for cognitive systems by looking at the extensions and adaptations made in three specialized models or frameworks. The first is Dick Hayes' and Linda Flower's cognitive model for expository writing. The second is the Card, Moran, and Newell models of human–computer interaction. The third is an architectural framework developed by our group that combines aspects of both the Hayes and Flower and the Card, Moran, and Newell models. In all three discussions, I emphasize the steps taken by these researchers that have enabled them to address these more complex tasks and situations.

Writing

As noted previously, I consider writing to be the quintessential conceptual construction task. First, writing requires a number of different intellectual skills used to transform an inchoate, loosely structured network of concepts into a well-structured, clearly expressed document. Thus, it is an inherently complex process that is made more so by extrinsic and document-specific factors — such as interruptions and the availability of necessary information. Second, the vast majority of tasks that involve building complex structures of ideas express those ideas as some form of document. Thus, writing is an integral part of most knowledge-construction tasks. Third, at a sufficiently abstract level, many of the processes and strategies used to plan, write, and revise documents can also be used to plan, express, and refine other types of information. This is equally true for collaborative as well as individual work. Thus, if we can understand the cognitive and social processes involved in writing, we will be well on our way to understanding a number of other tasks, as well.

During the past 15 years, a great deal of research has been done concerning the mental behavior of writers. The most important and most influential body of work is that of Dick Hayes and Linda Flower. In this section, I discuss their cognitive model of writing. In doing so, I want to emphasize two points: the IPS basis of their model, and, second, the adaptations and changes in research perspective they made that enabled them to address a task as ill-defined as writing.

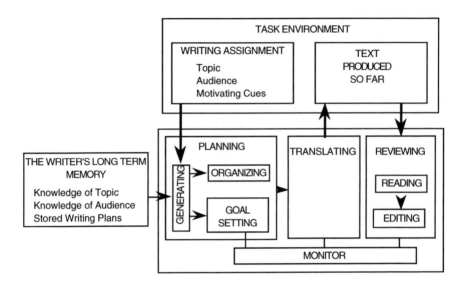

Fig. 4.4. Hayes and Flower's cognitive process model of writing. (Adapted by permission from Hayes & Flower, 1980.)

Hayes and Flower studied expository writing, as done by both students and adult professionals. Because they are colleagues at Carnegie Mellon University, it is not surprising that many of their methods and concepts were based on the earlier IPS model and the methodology developed by Newell and Simon to study human problem-solving behavior (Hayes & Flower, 1980; Flower & Hayes, 1984). For example, the Hayes and Flower writing model adopts a basic information processing perspective, and they asked writers to think aloud as they planned, wrote, and revised texts, just as Newell and Simon asked subjects to think aloud as they played chess or solved cryptarithmetic problems.

The Hayes and Flower model of writing is shown in Fig. 4.4. It includes a representation of the task environment that includes "the problem" to be solved — in this case, the writing assignment. It contains a long-term memory component that includes several types of semantic knowledge relevant to the task. The third major component of the model, shown at the lower right of the figure, is not labeled; presumably, it is a processor component because it contains three high-level processes and their associated subprocesses. Planning

GENERATING

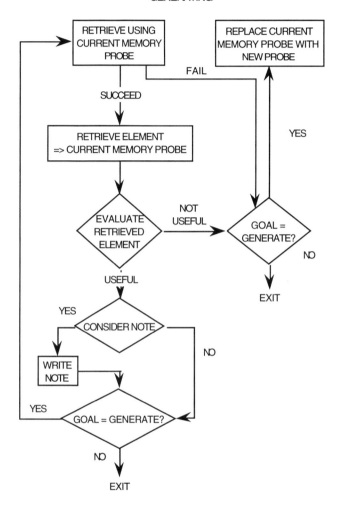

Fig. 4.5. Hayes and Flower's Generating (sub)process. (Adapted by permission from Hayes & Flower, 1980.)

includes three subprocesses: memory access (generating), a process for constructing a plan for the document expressed in some schematic form (organizing), and a goal-setting component to control movement among these processes. Translating, which contains no subprocesses, is concerned with encoding the ideas produced during planning into continuous prose — the text produced so far. Finally, reviewing involves two subprocesses: reading the text produced so far and editing it. The monitor is a different type of process, which I will discuss in just a moment.

Processes are represented in the model as decision procedures, expressed as flow charts. Fig. 4.5 shows the flow chart for generating, a subprocess that is part of the larger planning process. It is driven by three operations. First, long-term memory is accessed using the current contents of working memory as a cue. Second, an evaluation is made to determine whether or not the retrieved concept is useful. Third, if the concept is useful, it is externally represented as some form of note — Hayes and Flower's term for any type of information other than sustained prose, such as a word, phrase, symbol, and so forth. This (sub)process continues so long as goal = generate. When the goal changes, a new process is engaged.

Overall control of the writing system is provided by the monitor. It engages the different processes in accord with a set of strategies using a goal/subgoal mechanism, similar to that found in general IPS models and architectures. It differs from those systems, however, in storing only the current goal, rather than a hierarchy of goals/subgoals. The control program that runs in the Monitor is shown in Fig. 4.6. It is defined as a sequence of production rules. All decisions are based on the current contents of working memory. When working memory contains continuous prose, the monitor engages the edit process (rule 1). When it contains abstract information, the generate process is engaged (rule 2). When it simply contains a goal, the corresponding process used to achieve that goal is engaged (rules 7–10). Goal-setting productions that change the current goal (acted on in rules 7–10) appear as rules 3–6 in the program. Here, individual differences among writers come into play. Hayes and Flower describe four different strategies found in the writers they studied. Each strategy, which they refer to in the figure as a configuration, consists of an alternative sequence of rules 3–6. These alternative strategies are inserted into the overall control sequence in accord with the individual writer's personal writing habits.

Hayes and Flower used their model as a framework with which to analyze specific writing behaviors. They typically asked writers to think aloud as they plan and write their documents. Human judges then code the individual statements in the transcribed protocol to identify which cognitive (or metacognitive) process is currently

```
 1.   [Generated language in STM → edit]
 2.   [New information in STM → generate]
 3.-6. Goal setting productions (These vary from writer to
       writer; see different configurations, below.)
 7.   [(goal=generate) → generate]
 8.   [(goal=organize) → organize]
 9.   [(goal=translate)→ translate]
10.   [(goal=review) → review]
```

Configuration 1 (Depth first)
```
 3.   [New element from translate      → (goal=review)]
 4.   [New element from organize       → (goal=translate)]
 5.   [New element from generate       → (goal=organize)]
 6.   [Not enough material             → (goal=generate)]
```

Configuration 2 (Get it down as you think of it, then review)
```
 3.   [New element from generate       → (goal=organize)]
 4.   [New element from organize       → (goal=translate)]
 5.   [Not enough material             → (goal=generate)]
 6.   [Enough material                 → (goal=review)]
```

Configuration 3 (Perfect first draft)
```
 3.   [Not enough material             → (goal=generate)]
 4.   [Enough material, plan not complete→ (goal=organize)]
 5.   [New element from translate      → (goal=review)]
 6.   [Plan complete                   → (goal=translate)]
```

Configuration 4 (Breadth first)
```
 3.   [Not enough material             → (goal=generate)]
 4.   [Enough material, plan not complete→ (goal=organize)]
 5.   [Plan complete                   → (goal=translate)]
 6.   [Translation complete            → (goal=review)]
```

Fig. 4.6. Hayes and Flower's monitor control program, with four different strategies used by individual writers. (Adapted by permission from Hayes & Flower, 1980.)

operating. Figure 4.7 shows a single writing session for a subject analyzed in this way. Each think aloud statement is shown at the time it occurred and classified as one of four processes. The session can be divided into three large sections in which planning, writing, and revising processes dominate, respectively. Using analytic diagrams such as this, Hayes and Flower were able to manually trace the writing behaviors of their subjects through the various levels of their model, confirming that the model could account for the strategies and shifts from process to process exhibited by them.

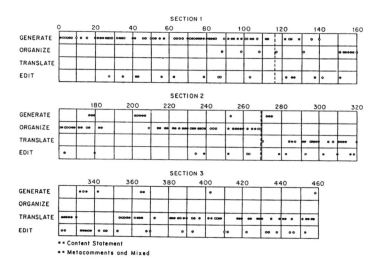

Fig. 4.7. Sequence of processes generated by one subject during a writing session. (Adapted by permission from Hayes & Flower, 1980.)

Such analyses led to a number of insights into the writing process that were not incorporated into the model, per se. For example, in a study of the differences in strategies between expert and novice writers, they found that the goals structures generated by the experts were so much more extensive and complex than those of the novices, the two groups of writers could be viewed as carrying out different tasks, rather than exhibiting different strategies for the same task (Hayes & Flower, 1986).

In summary, the Hayes and Flower model has much of the look and feel of the formal IPS models and architectures discussed in the preceding section. It consists of three major types of components: the overall framework for the model, represented as a box structure; specifications for individual processes, represented as flow charts; and a control procedure, represented as production rules, that includes alternative strategies. It describes both the overall writing process common to all writers as well as specific strategies used by individual writers, and a human interpreter can trace the behavior of an individual writer through the model. Research that has used the model as a framework for analysis has led to a number of interesting and useful insights that have enriched the model without being incorporated into it.

Although the Hayes and Flower model offered a much more detailed view of the writing process from what existed prior to its publication, one must keep in mind its limitations. The model is not sufficiently precise that it could be implemented as a simulation program or that it could be used to generate strong predictions of writers' behaviors. This is not a deficiency but, rather, indicates a different set of assumptions and goals. It was intended to be an analytic and conceptual model, to guide studies of individual writing behaviors, and to help researchers understand this particularly complex intellectual process. It took this approach to research because it promised more interesting and more practical results, given current knowledge of the task. And it has delivered good results. At some future time, using what we learn from this and other similar analytic models, we may eventually be able to build interesting predictive models or simulation systems for writing, but that time seems distant.

Finally, granting Hayes and Flower the informality of their approach, basic components of the model need to be defined more precisely. For example, process is not defined explicitly but, rather, by example through descriptions of specific processes. If we step back and ponder their underlying form, processes appear to be large-grain structures similar to problem spaces, as opposed to the small-grain cognitive operations or methods that occur within those spaces in other IPS systems. Each process is associated with a goal that causes the monitor to invoke that process. Processes include subprocesses. And they normally produce some form of intellectual product, such as a note that represents a concept retrieved from long-term memory, or a change in an existing product, such as an editorial change made to the Text Produced So Far. However, both the data types and structures used within processes are left as informal concepts, such as the traditional notion of a text, rather than as formally defined components of the model. Thus, although processes seem very similar to problem spaces, the relationship between the two is not spelled out. What is needed is an additional level of detail that makes explicit the basic architecture on which the model is built.

Human–Computer   Interaction

A second application of IPS concepts to a specialized task or situation is models of human–computer interaction. Just as writing is a part of many collaboration tasks, so an increasing number of groups use computer systems to help them with their work. Consequently, this work is also highly relevant for a concept of collective intelligence. In this section, I describe a particular view of human–computer interaction (HCI) research that has dominated the field for nearly a decade.

Card, Moran, and Newell (1983) defined a general framework in which specific models can be developed that characterize the behaviors of users performing specific tasks with specific computer systems. The framework has two major parts. Their model human processor (MHP) provides both a cognitive architecture and a set of quantitative measures derived from the literature for a range of basic processes and motor actions required to work with a computer. The second part of the framework is a set of categories for describing computer-related tasks. Because Allen Newell was one of the co-authors, it is not surprising that this framework resembles both the earlier Newell and Simon IPS model and Newell's more recent Soar architecture.

The model human processor, shown in Fig. 4.8, includes the following architectural components: a long-term memory, a working memory that includes separate stores for visual and auditory information, perceptual and motor processors, and a cognitive processor. A key aspect of this model is the set of parameters derived from the literature that quantize both the capacity of components (e.g., the amount of information that can be held in working memory) and the time required for fine-grained cognitive actions (e.g., the amount of time required to access long-term memory or to move the eyes). Thus, if one can decompose a task into a sequence of basic cognitive and motor actions, then one can use the model to predict the time that will be required for someone to perform that task.

The basic MHP model is extended to include some 10 higher level regularities, called principles of operations. Two examples are the *power law of practice*, which describes the time expected for a user to perform a given task after a varying number of practice trials, and *Fitt's law*, which describes the relationship between the time required

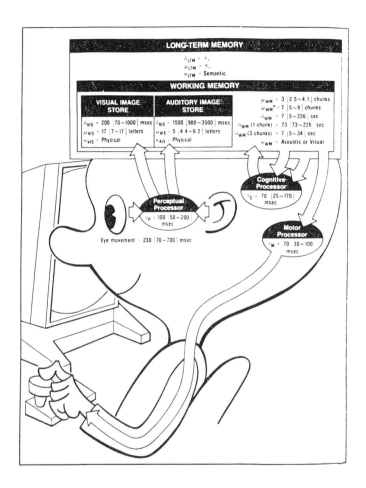

Fig. 4.8. Card, Moran, and Newell's Model Human Processor. (Reproduced by permission from Card, Moran, & Newell, 1983.)

by a user to move his or her hand a given distance toward a target and the size of that target. These regularities provide larger behavioral structures in which to fit the more basic processes and components of the MHP architecture.

The second major part of the Card, Moran, and Newell framework is a set of categories, called GOMS, that is used to define models for tasks performed with a given computer system. GOMS is an acronym for *goals, operators, methods,* and *selection rules.* A task typical of those that have been described in terms of GOMS categories is a user of a particular computer editor making changes to a document from a previously marked copy of the text (Card, Moran, & Newell, 1983). Tasks are described in terms of a set of goals. Higher level goals generate lower level goals; for example, a goal of making a given change in the computer version of the text as indicated in the marked paper version might generate the subgoal to find the appropriate location in the computer version where the change is to be made. Thus, tasks produce hierarchies of goals and subgoals, similar to those produced in the other IPS models and architectures discussed above. Operators are basic system actions, such as a menu selection or a single typed command. They represent the user's knowledge of how the system works. Methods are short sequences of operators known to the user that are routinely used to accomplish a goal. Thus, for example, inserting a word in a line may require moving the cursor, typing a control sequence, typing the character string to be inserted, followed by typing another control sequence to conclude the task. Sequences of operations such as these may be used so often that they become automatic.

The most complex part of the GOMS model is the set of selection rules. Normally written as production rules, they identify a set of conditions and the action that will follow when those conditions are met. Thus, they predict users' behaviors under different conditions. Because goals may be part of both the conditional and action parts of a selection rule, they can generate hierarchies of goals and subgoals that eventually terminate in specific methods defined as sequences of basic operations. Consequently, by applying the timing parameters from the model human processor model, one can also predict the amount of time these sequences will take, and hence the task as a whole.

The selection rule component has been extremely influential in determining the kinds of problems GOMS can and cannot address. GOMS has been used most successfully for predicting short, independent sequences of actions, ranging from three or four to a dozen or so. Tasks comprised of sequences of this length are usually measured in seconds or tens of seconds. Typical examples, in addition to the editing task described previously, include performing independent operations with a spreadsheet program (Olson & Nilsen, 1988) and playing video games (John & Vera, 1992).

Two recent extensions to the basic GOMS model have extended this limit. The first, called CPM-GOMS, includes a critical path component and has been used to model the behavior of long distance telephone operators (Gray, John, & Atwood, 1992). It enabled researchers to predict behaviors of operators using two different computer workstations for transactions involving 15 to 20 operations and lasting about as many seconds. A second extension, called Browser-Soar, combines the basic GOMS approach with the problem space concept from Soar. Developed by Virginia Peck and Bonnie John at CMU, Browser-Soar has been used to model independent browsing operations performed by users of a help system lasting for approximately 100 seconds and involving several subtasks (Peck & John, 1992).

Although CPM-GOMS and Browser-Soar significantly extend the GOMS approach, the tasks they model are quite different from the sustained, interdependent activities that occur in knowledge-construction tasks, such as writing. Consider what would be involved in developing a detailed GOMS model for expository writing, comparable to the Hayes and Flower model. This would require predicting the order in which a writer will engage the different processes identified in the model: when and under which conditions he or she will switch from one process to another; when a particular line of thought will run dry; when someone will hear a stray conversation in the background and be reminded of an idea that can be incorporated into the document; and so on. A predictive model that could handle behaviors of this sort would constitute an artificial intelligence capable of simulating expository writing. This is not to say that all aspects of complex behavior are unpredictable. Indeed, we can predict behaviors such as the average length of time spent in continuous work episodes based on attention span and limits of working memory. But predicting specific cognitive behaviors that are affected by complex strategies and tactics, the semantics of the task domain, and extrinsic factors lies far beyond current capabilities.

This difference between a GOMS approach and a Hayes and Flower approach is a paradigmatic difference. The first is quantitative and formal and assumes that the ultimate value of a model, as well as its verification and refinement, are based on its capability to generate predictions. The second is qualitative and informal and assumes that the value of a model lies in its capability to provide interesting and useful insights into complex, real-world tasks. The second does not deny the value of the first; it just defers it until such future time when the base of knowledge is sufficient that generative models can be developed that can handle the same factors and situations. Although the Hayes and Flower model was not entirely satisfactory, it illustrates the kind of model that is likely to be most practical and most useful for current considerations of collaborative groups.

Although the MHP/GOMS framework is based on IPS concepts, it is not as developed. Like other IPS systems, it includes goals, goal/subgoal hierarchies, and a set of basic operations. These components are situated within the user's long-term and working memories. But the framework does not include the additional structure provided by a problem space. Thus, it includes no inherent concept of data type or data structure. Such concepts may be part of the computer system presumed by a GOMS model, but direct references to conceptual data must be incorporated into other components of the framework. Frequently, goals have served this purpose. But this overloads the concept: Sometimes goals refer to abstract intentions, at other times to changes in the system's data model that will realize those intentions. Given Soar's subsequent inclusion of both a problem space and its associated functions as basic components of the human cognitive architecture, omitting some form of larger context in which cognition is presumed to take place appears to be a severe limitation in GOMS.

Finally, the MHP/GOMS approach has been used most frequently to characterize the behavior of users working with existing computer systems. It has not been used extensively as a tool for designing new or improved systems. One reason for this is that there is no independent cognitive component in the framework. As a result, task descriptions are normally cast in terms of the options offered by an existing computer system — operators or sequences of operators — for accomplishing a basic task goal. Consequently, there is no natural mechanism in which to develop a separate cognitive model for a task that can then be mapped onto a separate system model that, in turn, can inform subsequent design and implementation of that system. If we wish to develop computer systems that closely match users' cognitive behaviors and strategies, as is the goal for *intelligence amplification* systems, we will need capabilities that go beyond those included in the MHP/GOMS framework.

Cognitive Modes and Strategies

In this section, I discuss a framework developed by our research group that can be used both to describe complex, real-world knowledge-construction tasks and to develop theory-based computer systems to support those tasks. Its basic constituents are a set of *cognitive modes* used by individuals to perform a given task and the *strategies* they use to guide them as they shift from one mode to another in carrying out that task.

The approach combines aspects of both the Hayes and Flower and the Card, Moran, and Newell models/frameworks. Like the Hayes and Flower approach, the mode/strategy approach is oriented toward analysis and description, rather than prediction; thus, it, also, bypasses the limiting factor in the GOMS approach — selection rules — by regarding users' strategies as the object of discovery rather than a constituent that must be in place prior to analysis. Consequently, mode-based models can address coherent, interdependent behaviors that extend over hours, days, and even longer periods of time. Like the Card, Moran, and Newell approach, the mode/strategy framework is a well-defined architectural construct; thus, it avoids the ambiguity of the Hayes and Flower process model. However, unlike MHP/GOMS models, mode/strategy models can be mapped directly onto system design. Thus, the mode/strategy framework retains key features of these other approaches while addressing several of their limitations.

In this section, I first define the mode/strategy framework as a general architectural construct and then illustrate its use by describing a set of modes for expository writing. After that, I describe a writing support system whose design was based upon that set of modes. Finally, I describe a mode-based analytic model developed by our group for studying the cognitive strategies of writers using the system.

Mode/Strategy Framework

A *cognitive mode* is a particular way of thinking used for a particular purpose. For complex tasks, such as conceptual construction tasks, human beings engage different cognitive modes in order to accomplish different parts of the task. Consequently, they move from one mode to another in accord with strategies they know for the task and in response to changing conditions in both the content domain and the external environment.

A mode is determined by four factors: goals, products, processes, and constraints. Thus, a particular mode of thought is associated with a particular goal that will be realized by producing a particular type of conceptual product, drawing on particular cognitive processes in accord with a particular set of constraints (Smith & Lansman, 1989). By implication, a given mode may exclude or discourage certain kinds of products and the processes used to develop those products. Let's look more closely at each of these four factors that identify a given mode. Although the discussion is general, I use expository writing as the example task. Other tasks would include other sets of modes, but those modes would be cast within the general framework described here.

Goals represent the intentions that lie behind a person's use of a given mode. Thus, for example, one may engage a brainstorming mode as part of the overall task of writing in order to gain a general sense of the information that is available to the writer. A goal is normally realized by creating some form of conceptual product or by making a change in one or more existing products, but the two are not the same: Goals are abstract; products are concrete.

Different cognitive modes provide different options for representing concepts or structures of concepts. Consequently, different modes support the development of different types of information *products*. These include words, phrases, sentences, outlines, diagrams and drawings, symbols, equations, computer code, and other forms. Thus, product encompasses both the data type and the data structure components of the problem space architecture.

Cognitive *processes* act on cognitive products to create them, to extend or modify them, or to transform one type into another. For example, some processes are concerned with accessing and representing a portion of the associative or semantic network of an individual's long-term memory. Others are concerned with transforming that network into a hierarchy. Still others are concerned with translating abstract ideas into specific representations, such as sentences or diagrams. Consequently, certain processes are favored in particular modes, whereas other processes are de-emphasized or even suppressed. Thus, process is a small-grain concept in the modes framework, comparable to operator in GOMS and IPS systems, but distinctly smaller than the process found in the Hayes and Flower model.

Constraints determine the choices available within a mode. They set thresholds on evaluation functions that guide the flow of conceptual thought. Examples of these functions, which usually operate automatically, include determining the relevance of a retrieved concept, evaluating a change made to a conceptual structure, and invoking or suppressing corrective processes. Constraints are raised or lowered in different modes in accord with the general goal or purpose for engaging that particular way of thinking. Thus, they set the overall "tone" for a mode. None of the systems discussed previously include constraints as a basic component of the architecture, although a limited form of constraint is implicit in the evaluative component of basic strategies, such as means–ends analysis, used in IPS problem spaces.

To gain a better feel for the way interdependent combinations of these four factors determine particular modes of thought, let's look more closely at two specific modes used by many writers: *exploration* and *organizing*. During exploration, the goal is to externalize ideas and to examine those ideas in different combinations and relationships with one another. Consequently, constraints are kept to a minimum to encourage creativity and multiple perspectives. For example, during this mode, many writers pay little attention to spelling, neatness, or syntactic precision; they may even suppress their tendencies to notice these things. They may also try not to make decisions about which ideas are relevant or not relevant in order to generate a larger pool of possibilities from which to eventually select. As a result, the products generated are often informal, consisting of notes, jottings, diagrams, loose networks of concepts, and so on. Processes include recalling concepts from memory, basic encoding, associating and relating concepts, and building small component structures.

During organizing, the goal is to plan the actual document to be written. Consequently, constraints are tightened and thinking becomes much more rigorous and systematic. The goal is usually achieved by creating some concrete representation of a plan for the document, such as an outline, tree, or other form of hierarchical structure. The processes emphasized in this mode are those needed to construct the plan. They include analyzing; synthesizing; noting various relationships between ideas, such as cause and effect or subordinate–superordinate relationships; and comparing different parts of the evolving plan for consistency and parallel structure.

Thus, exploration and organizing are distinctly different ways of thinking. And they differ from other activities such as translating the abstract ideas in the plan into sentences or editing the resulting document. Figure 4.9 provides a more complete set of modes for expository writing. In addition to exploration and organizing, it

	Processes	Products	Goals	Constraints
Exploration	•Recalling •Representing •Clustering •Associating •Noting subordinate-superordinate relations	•Individual concepts •Clusters of concepts •Networks of related concepts	•Externalize ideas •Cluster related ideas •Gain general sense of available concepts •Consider various possible relations	•Flexible •Informal •Free expression
Situational Analysis	•Analyzing objectives •Selecting •Prioritizing •Analyzing audiences	•High-level summary statement •Prioritized list of readers (types) •List of (major) actions desired	•Clarify rhetorical intentions •Identify and rank potential readers •Identify major actions •Consolidate realization •Set high-level strategy for document	•Flexible •Extrinsic perspective
Organizing	•Analyzing •Synthesizing •Building abstract structure •Refining structure	•Hierarchy of concepts •Crafted labels	•Transform network of concepts into coherent hierarchy	•Rigorous •Consistent •Hierarchical •Not sustained prose
Writing	•Linguistic encoding	•Coherent prose	•Transform abstract representation of concepts and relations into prose	•Sustained expression •Not (necessarily) refined
Editing: Global Organization	•Noting large-scale relations •Correcting inconsistencies •Manipulating large structural components	•Refined text structure •Consistent structural cues	•Verify and revise large-scale organizational components	•Focus on large-scale features and components
Editing: Coherence Relations	•Noting coherence relations between sentences and paragraphs •Restructuring to make relations coherent	•Refined paragraphs and sentences •Coherent logical relations between sentences and paragraphs	•Verify and revise coherence relations within intermediate sized components	•Focus on structural relations among sentences and paragraphs •Rigorous logical and structural thinking
Editing: Expression	•Reading •Linguistic analysis, transformation, & encoding	•Refined prose	•Verify and revise text of document	•Focus on expression •Close attention to linguistic detail

Fig. 4.9.   Seven cognitive modes for expository writing, including the processes, products, goals, and constraints for each mode.

includes *situational analysis*, in which the rhetorical context of the document is explored; *writing*, per se; and three modes for *editing*: *organizational* editing, involving large structural components of the document; *coherence* editing, involving relationships within individual paragraphs or small sections of the document; and *expression* editing, involving individual sentences (or other basic types of information).

The preceding discussion described cognitive modes as separate "islands" of thought to emphasize their distinctness. However, modes are also related to one another in fundamental ways. First, individuals shift from one mode to another over time; consequently, they exhibit different tendencies or patterns of behavior in the order in which they engage different modes and the conditions that cause them to shift from one mode to another. Second, the intellectual products created in one mode are often carried to or appear in another mode in which a different set of processes is used to continue development or to transform one type of product into another.

The first relationship is concerned with the *strategies* and *tactics* an individual uses to accomplish a task. Strategy refers to an individual's overall understanding or image of a task and the large-grained process that person has learned or developed that enables him or her to accomplish that task. Examples of strategy include the "stages" model of writing and the "waterfall" model of software development. Tactics refer to the shifts people make from one mode to another in order to respond to problems that arise or to changes in conditions. For example, writers may return to organizing mode when they realize during writing that the plans they constructed earlier have problems (Hayes & Flower, 1980). Thus, modes help individuals focus their attention on a single activity at a time, whereas strategies and tactics provide them with the means to move from one activity to another in coherent ways. Of course, not everyone uses the same strategy for a given task; in fact, differences in individual behavior can be characterized in terms of patterns in the sequences of modes they engage (Lansman & Smith, 1993).

The second relationship is concerned with the transfer of information from one mode to another. When an individual follows a global strategy for a task, he or she normally produce a *flow* of intermediate products in which the output of one mode becomes the input for another. For example, during exploration, many writers represent concepts externally, cluster them, and then link them into a loose network of associations. During organizing, they transform this loose network into a coherent, consistent structure for the document.

If they use an outline or tree for this purpose, then the transformation is from a network to a hierarchy.   During writing, they transform abstract concepts and relations in the plan into continuous prose, graphic images, or other types of information.   During editing, they refine the structure and expression of a draft document to produce an improved or final version.

This flow of products, however, is not one-way and continuous. As an individual shifts from one mode to another, intermediate products flow back and forth, as well.   For example, writers may find while organizing that they do not have crucial information needed for a particular section.   Rather than interrupt their thinking to seek out that information then, they may decide to continue organizing but leave the relevant section undeveloped.   Later, when the missing information is available, they may interrupt their writing, revert to organizing and/or exploration modes to build the missing portion of the document's structure.   When the missing part has been filled in, they resume writing (Smith & Lansman, 1991).

In summary, for many intellectual activities, individuals divide tasks into subtasks, set goals and subgoals, produce intermediate as well as target products, and employ different processes to produce them.   They use general strategies to guide overall behavior and more specific tactics to resolve problems.   The behavior of particular individuals or groups carrying out specific tasks can be modeled by identifying the particular set of cognitive modes they use along with their strategies and tactics.   Thus, the general concepts of *mode, strategies,* and *tactics* can be viewed as an architectural framework that can be applied to a broad range of tasks.

Mode-Based Writing Environment

If a task is described in terms of a set of cognitive modes and the expected flow of intermediate products from one mode to another, that description can be used in a direct and natural way to guide design of a computer system to support that task. This is done by including different working contexts for the different cognitive modes and providing mechanisms for moving data from one context to another. (Smith & Lansman, 1992).

More specifically, after identifying a set of cognitive modes for the task, one can then design a corresponding set of interface or *system modes*. Each system mode can be associated with a different window in the user interface. Each supports a different data model that includes both data types and data structures appropriate for its corresponding cognitive mode(s). Each system mode should also include a set of functions sufficient to construct data objects of that type. Ideally, these functions would be presented in one-to-one relationship with corresponding processes in the cognitive mode; however, in some cases, several system operations may be required to represent the results of a single cognitive process. Thus, in general, there is a many-one relationship between cognitive process and system actions.

To support the flow of intermediate products from one cognitive mode to another, the system should provide functions for copying or moving data from one system mode to another. An alternative design would be to overlay the data with a succession of modes, rather than to move data between modes; however, this approach is neither as general nor as flexible as the flow model and can lead to problems, such as blocking shifts between pairs of modes that have distinctly different data types/structures.

An example system whose design was based on a mode/strategy task model is the Writing Environment (WE). Built by our research group, WE includes four system modes that support six of the seven cognitive modes included in Fig. 4.9 (Smith et al., 1987). A sample screen for WE is shown in Fig. 4.10.

Network mode, shown in the upper left window, supports exploration. The underlying data model is a directed graph embedded in a two-dimensional space. Thus, the user has maximum flexibility for representing concepts as nodes (boxes with a word or phrase to express the idea), moving them to form clusters of loosely related ideas, and linking them to denote more specific relationships. Small conceptual structures can also be built here and later used in other modes. Although writers may do all of their large-grain structural work in this mode, the system provides another mode that is better suited for building the actual plan for the document.

Tree mode, shown in the lower left window, supports the organizing cognitive mode. The underlying data model is a tree or hierarchy, and the functions provided for building the tree are

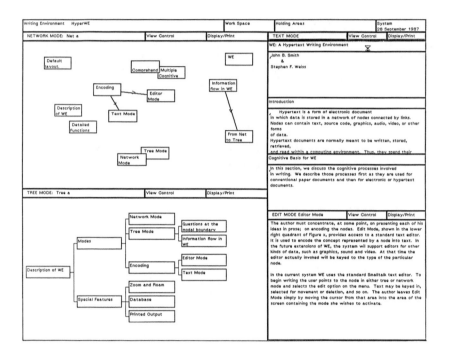

Fig. 4.10. Sample screen for the Writing Environment (WE), showing network, tree, editor, and text system modes.

constrained so that one cannot denote a relationship that would violate the integrity of the hierarchical structure. Thus, the functions and constraints of tree mode are different from those of network mode in which maximum flexibility is emphasized. This design decision was based on research in reading comprehension that shows that, in general, hierarchical documents are more easily and more accurately comprehended than unstructured documents or documents with other structures (Smith & Lansman, 1989). Although users may create the tree from scratch, more often they combine creating new nodes in tree mode with copying nodes or small component structures from network mode. Thus, system design encourages writers to transform the loosely structured network of ideas, developed in exploration/network modes, into consistent, well-defined hierarchical structures in organization/tree modes. But it does not require them to do so.

The writing cognitive mode is supported by *editor* mode, shown in the lower right corner. At any time in the overall writing process, users can open a node in either the network or tree modes and write a block of text in the editor that will be associated with that node. Because writers work on the contents of a single node at a time, the design of the system encourages them to focus their attention on a single concept and to transform that abstraction into linguistic expression. Thus, at any given time, the writer is concerned with writing a small, manageable component as opposed to writing the entire document, thereby simplifying the overall writing/translating task. To construct a conventional linear document from all these pieces, the system "walks" the tree — top to bottom, left to right — and gathers up the contents of the various nodes. It then sends these concatenated pieces to a file or printer.

Finally, the upper right system mode, called *text* mode, is intended for coherence editing. Although the system reduces users' cognitive load by enabling them to work with the structure of the document in abstract, schematic form (tree or network) and to divide the actual writing of the document into a succession of small writing subtasks, this approach can lead to documents with inconsistencies and awkward transitions between sections. To address these problems, text mode presents the document in its linear, concatenated form for editing. Thus, one can read the text continuously "across" node boundaries, check transitions, move sentences from one node to another, and so forth.

Thus, WE's four system modes — network, tree, editor, and text — correspond to four cognitive modes — exploration, organizing, writing, and coherence editing. For organizational editing, writers use tree mode — by moving branches and nodes around in the tree, they can reorganize the text of the associated document. To support expression editing, writers use either editor or text modes. Consequently, six of the seven cognitive modes described in Fig. 4.9 are supported by the four WE system modes. WE does not support situational analysis mode in which writers analyze the rhetorical context and make strategic decisions about their documents; however, we have developed heuristics to assist with this process (Smith & Smith, 1987), which could be incorporated into future versions of the system.

Ideally, development of the task model precedes system design. In practice, development of the two may be an iterative process, with work in down-stream phases — for example, system building — informing work in up-stream phases — for example, model building. Ultimately, it is not important which came first, but that the two end up being consistent with one another. The mode/strategy framework can contribute to this process by making the relationships between the two straightforward.

Mode-Based Analytic Model

In the preceding sections, I showed, first, that knowledge-construction tasks, such as expository writing, can be described in terms of a set of cognitive modes, the transitions that occur between modes, and the resulting flow of intermediate products from one mode to another; and, second, that mode-based task descriptions can be mapped onto system design in a natural and straight-forward way. In this section, I take this synthesis one step further by introducing the notion of an analytic model that takes into account the mediating effects of such a system on task behavior. Here, I only introduce the idea in order to complete this picture of the mode/strategy approach; in chapter 7, I describe several such models in more detail in discussing the general concept of *strategy*.

In order to study the cognitive behaviors of individuals using the WE system to plan and write documents, we instrumented the system so that each time a user selected a data object, chose a menu option, or moved from one system mode to another, a record of that action was recorded. The sequence of all such records for a session comprise what we call an *action level protocol*. These data provide a detailed record of the system functions used to represent the results of users' cognitive processes. They also identify cognitive products produced by those processes and their evolution over the course of the task. Thus, they provide a record of the material production of a document, from earliest brainstorming and planning to final editing. However, because not all thought results in a system action, we cannot claim that the record is complete or that it includes no distortions, but because system model and task model are closely related, the action protocol should represent to an approximation a trace of the user's thought process during the task.

Analytic models can be developed to analyze these protocol data and to uncover patterns in users' cognitive behavior. We chose to express these models as grammars. A conventional grammar takes as input a string and determines whether or not that string is contained in a language. Thus, for a natural language such as English, a grammar takes as input a string of words and determines whether they constitute a valid sentence in English. To do so, the grammar produces a parse of the sentence that shows its grammatical structure. An analytic model that is expressed as a grammar takes as input an action level protocol and produces as output a parse tree that describes the user's strategy for the session.

We developed several models that can analyze writers' strategies. One produces parse trees that include six hierarchical levels (Smith, Rooks, & Ferguson, 1989). The root of the tree is the *session*. Each session is divided into a sequence of cognitive *modes*. Each mode, in turn, is divided into a sequence of cognitive *processes*. Each process symbol is then linked to one or more cognitive *products*. Thus, the top four levels of the model are cognitive and, hence, independent of the WE system.

The bottom two levels of the grammar map the cognitive portion of the model onto the design of a specific computer system — in this case, WE. To represent a change to a cognitive product, users perform one or more system *operations*, each of which requires several system *actions*. Because each protocol record corresponds to a single system action, actions are the terminal symbols in the grammar, analogous to individual words.

This grammar is expressed as a set of production rules, supplemented by some half-dozen functions that recognize particular context-sensitive relationships. It is implemented as a computer program; thus, it can be applied consistently and automatically across multiple user protocols. We used the parser to analyze a number of user sessions under different experimental conditions. Results of these studies are described in Smith and Lansman (1989), Lansman (1991), Smith and Lansman (1992) and Lansman and Smith, (1993).

Perspective

Like the IPS architectures discussed previously, the mode/strategy framework provides an architectural construct that applies to a wide variety of tasks. It includes an explicit context in which cognition takes place as well as a set of components that occur or operate in that context. It differs from them, however, by including constraints and a richer set of product types as basic components and by regarding strategies and tactics as the primary objects of discovery.

The mode/strategy framework encourages development of analytic models for complex, real-world conceptual construction tasks, rather than predictive models or simulation systems for simple or artificial tasks. In this respect, it is similar to the Hayes and Flower approach and differs from the GOMS and IPS approaches that emphasize generative models and simulation systems. The reasons for this are practical — we simply do not have an adequate base of knowledge at this time to support generative models that can address interesting or meaningful issues for this category of tasks. However, by developing formal analytic models and by using those models to study users' strategies and behaviors, we may eventually be able to build a sufficient base of knowledge that would make generative models of complex tasks practical. Thus, I see the two approaches as potentially complementary.

Like GOMS, the mode/strategy framework can be used to describe users' interactions with computer systems. However, unlike GOMS, it provides direct guidance for designing new or improved systems. This results from the direct and natural mapping between a set of cognitive modes and the flow of information among them to a set of system modes and the flow of data among them. Because a system built in this way has a well-defined relationship with a cognitive model of the task, we can identify the specific ways in which that system amplifies its users' cognitive skills as well as the extent of change. Thus, the mode/strategy framework makes it possible to talk about intelligence amplification in precise and meaningful ways.

Although the mode/strategy model of writing resembles the Hayes and Flower model in its emphasis on analysis, it differs from it in several important respects. First, it is defined within the terms of an explicit architecture, rather than by example in terms of a collection of specific processes. Second, because it is expressed as a computer program, all of its terms and rules are well-defined. Thus, it is a formal, executable model, as opposed to an informal conceptual model. Third, it can be applied automatically and consistently across multiple subjects, rather than manually by human judges who sometimes disagree in their interpretations.

Many of these differences are subtle, but together they add up to an approach that is particularly well-suited for handling complex, computer-based conceptual construction tasks.

Objection to Collective Intelligence

Before concluding this discussion of concepts drawn from the cognitive science literature, I briefly note an objection raised by Allen Newell to the notion of a collective intelligence, based on the rate at which knowledge can be communicated from one individual to another. Because Newell has been so influential in the development of cognitive science, this objection must be addressed; consequently, I include his argument in his own words:

> A social system, whether a small group or a large formal organization, ceases to act, even approximately, as a single rational agent. Both the knowledge and the goals are distributed and cannot be fully brought to bear in any substantial way on any particular decision. This failure is guaranteed by the very small communication bandwidth between humans compared with the large amount of knowledge available in each human's head. . . . Modeling groups as if they had a group mind is too far from the truth to be a useful scientific approximation very often. (Newell, 1990, pp. 490-491))

Elsewhere, he explained in more detail his assumptions and reasoning regarding fundamental limits in human communication:

> Let the rate of knowledge intake (or outflow) between the human and the environment be $\sim\sim K$ chunks/sec. Then this same rate governs the acquisition of knowledge prior to the meeting of the group that attempts to share knowledge by communication. The total body of knowledge in a group member is $\sim\sim KT$, where T is the total lifetime of the member. The amount that can be shared in the group meeting is $\sim\sim K\Delta T$, where ΔT is the duration of the group meeting. But the group meeting time, ΔT, is small compared to T, independent of K, the communication rate. . . . The net effect is that the social band becomes characterized as a distributed set of intendedly rational agents, in which each agent has a large body of knowledge relative to how fast it can communicate it. (Newell, 1990, p. 155)

Thus, Newell argued that for a group to exhibit collective intelligence, all members of the group must share the complete body of knowledge and goals relevant to the task that are held separately by its individual members. Newell is right that no group can achieve total integration of knowledge such as this. However, this may be too strong a requirement. Individuals do not always utilize all of the potentially relevant knowledge they possess for each decision or mental operation they perform. Similarly, total shared knowledge may not be required to achieve a level of coherence within a group sufficient to justify a concept of collective intelligence. I try to speak to this objection at several points in the discussion that follows.

Summary

In this chapter, I looked at cognition as an information processing activity. The discussion began with a review of three general models or architectures developed from this perspective. A basic set of components was identified in Newell and Simon's original IPS architecture that are subsequently refined in Newell's Soar and Anderson's Act* systems. These include a long-term memory, possibly divided into separate stores for procedural and declarative knowledge; a working memory; and a separate processor component or, alternatively, a set of processes that function within working memory. These systems also included a context for higher level conceptual thought called a problem space. This context, in turn, included a goal, a data type, and a data structure as well as set of rudimentary strategies for constructing a path from an initial data state to the goal data state. The architectures differed in their respective views of a problem space as part of the basic cognitive architecture or as an informational object acquired as part of learning a particular skill.

Next, I looked at several models and frameworks that applied these ideas to specific tasks, such as writing, or specific situations, such as users working with computers to perform a task. These discussions added several new concepts. First, the process model of Hayes and Flower showed that by taking an analytic approach, one can address conceptual construction tasks that cannot currently be addressed using predictive or generative models. Second, the GOMS model provided a set of categories for describing tasks performed in close conjunction with a computer system. It is limited in the tasks it can model, however, by the fundamental role played by its selection rules component. Finally, the mode/strategy framework combines aspects of both approaches. Its practice of expressing task models as grammars leads to formal analytic models based on a well-defined architectural construct. When these grammars are implemented as parsing programs, they can support extensive studies of users' cognitive behaviors, perhaps leading to more sophisticated concepts of strategies and tactics, currently missing from conventional IPS models.

I also reviewed an objection raised by Allen Newell to the notion that a group can function as a coherent intelligent agent. It states that for a group to function as an intelligent agent, all members would have to share complete knowledge of goals and the task domain, an impossible requirement because of bandwidth limitations.

In Part II, I draw on this research in building a concept of collective intelligence by identifying components within collaborative groups that are recognizable as extrapolations of basic IPS architectural components. I also try to build a path around Newell's objection to this concept.

Part II

Building a Concept

o f

Collective Intelligence

In Part II, I try to build a concept of collective intelligence as a form of computer-mediated collaboration. In doing so, I consider the group and the computer system it is using to support its work as a single abstract *system*, synthesized from two subsystems — one human, one technological. It is the behavior of this composite system that I will address. The critical issues, then, are the factors that enable this system to function and, under some conditions, to produce work that is coherent, internally consistent, and has intellectual integrity.

The strategy I follow is to build up a description of this system piece by piece. Because I am working from an information processing perspective, those pieces are the primary components found in the IPS models and architectures discussed in chapter 4.

First, I consider *memory*. I examine constructs that function as a form of collective memory for the group. As might be expected, it includes provisions for the long-term storage and retrieval of information, analogous to human long-term memory, and it provides contexts in which that information can be activated and processed, analogous to working memory.

Second, I consider *processing*. The discussion focuses on the form and function of individual small-grained processes that operate in the contexts identified as the group's working memory. They are responsible for basic actions such as retrieving and defining concepts, identifying relationships, building conceptual structures, making changes to those structures, and storing results. Thus, they enable the group to function as an information processing system and to carry out its task.

Third, I consider *strategy*. The prior discussion of processes is limited to their basic, architectural forms, illustrated by specific processes found in collaborative groups.   Thus, processes are considered as independent entities or sets of entities. In the discussion of strategy, they are considered in relation to one another. The focus is on patterns in the sequences of processes that occur in the behavior of a group. By engaging processes not as isolated actions but in coherent sequences, groups are able to function in purposeful ways and to accomplish goals. Thus, the discussion moves to issues of collaborative problem-solving and knowledge-construction.

Fourth, I consider two metacognitive issues — *awareness* and *control*. Awareness plays a large role in enabling an individual to produce intellectual products that are coherent, internally consistent, and, occasionally, elegant. How can a group produce work with similar characteristics if its products are too large and too complex to be understood by any one person? To answer that question, I examine ways in which a group can piece together partial, but overlapping bodies of knowledge among its members to produce partial, but overlapping fields of awareness. The discussion also considers authority and administrative control within groups in relation to issues of self-control in individuals. A balance between delegated and centralized authority is needed to motivate groups and to enable them to produce work with intellectual integrity.

The discussion concludes by outlining a research agenda that could replace the *concept* of collective intelligence described here with a fully realized *theory*. Such a theory would enable detailed process models of collaboration that apply across multiple collaborative tasks in different social and organizational situations. Development of such models could lead to a number of practical benefits, including more effective training in collaboration skills, better support systems for distributed groups, and more productive organizational structures.

Chapter 5

Collective Memory

The cognitive models and architectures discussed in chapter 4 all included memory systems that provide both long-term storage of information and a context in which that information is processed. It seems self-evident that any system that attempts to model cognition as an information processing activity must include components or facilities that provide these two functions. If no long-term memory were included in the system, only current information would be available for processing, severely limiting any form of cognition that could occur in that system. If no working context were included, processing would have to occur within the input/output streams of the information flow. Although some sort of filter architecture might be possible, that architecture could not solve problems that require context sensitive operations or relating bits of information that are not adjacent in the input/output stream. Consequently, I assume that for a collaborative group to function as a coherent, intelligent system, that system must include components that provide long-term storage of information and contexts in which conceptual processing can take place. I refer to these components as the group's *collective memory*.

We might expect the collective memory to resemble human memory in broad outline. For example, it seems reasonable to assume that separate subsystems will be used for long-term storage and for active processing, that information will move back and forth between the two, and that individual operations will be responsible for retrieving concepts, changing their form and content, and storing the result. Otherwise, it is hard to imagine how groups can build complex artifacts. However, we should also expect the abstraction I am calling the collective memory to differ from human memory. At the very least, it must take into account the multiple working memories of the individuals who comprise the group. Consequently, the strategy I follow in this chapter is to define a construct that is recognizable as a *plausible* analog or extrapolation of human memory, which provides similar function, but is not necessarily identical to the corresponding components in conventional human memory system(s).

The information flow model introduced in chapter 2 included two basic types of information: tangible and intangible. Because we are considering collective intelligence and human cognition from an information processing perspective, we should consider the collective memory in relation to these types. That is, we must consider how tangible knowledge is stored over long periods of time and in which contexts it is accessed and processed. This results in memory subsystem for tangible knowledge. We must also consider similar issues for intangible knowledge, resulting in a second subsystem for that type of information. Together these two memory subsystems comprise the collective memory for a collaborative group.

Throughout this and the discussions that follow, it is important to recognize that we can view a group both as the collection of individuals who comprise it and as an entity in its own right. Consequently, when I speak of a group's collective memory, I am referring to a collective memory for that entity. It is related to the individual memory systems of the group's members, but it also functions with respect to the group as a whole.

The discussion is divided into two main parts. The first is concerned with components for storing and working with tangible forms of information, the second with comparable components for intangible knowledge. The chapter ends with a brief discussion of issues for further research. Because the focus is on collective intelligence in computer-based collaboration, I assume that groups are using a computer system, such as the ABC system described in chapter 3, to support their activities.

Tangible Knowledge

In this section, I focus on *tangible* knowledge and on constructs that provide functions similar to those provided by long-term, working memory, and extended memory in individuals. Together, these components can be regarded as the group's collective memory for tangible knowledge.

Long-Term Memory

Let's look first at long-term memory. One must keep in mind that the discussion is concerned with computer-based collaborative groups and with collective intelligence as a form of computer-mediated behavior. Because I am assuming that groups do most if not all of their tangible work in direct relation to a computer system, such as ABC, when we consider the long-term storage of the group's tangible knowledge, we can do so with regard to how that information is stored in the computer system.

I refer to the body of tangible knowledge that a group works with as the *artifact*, as suggested in Fig. 5.1. Within the ABC system, it is stored as a hypermedia graph structure. It is a straightforward extrapolation to regard the artifact as a form of long-term memory for the group. To see why this is so, we can examine the artifact with respect to important characteristics and functions associated with conventional long-term memory.

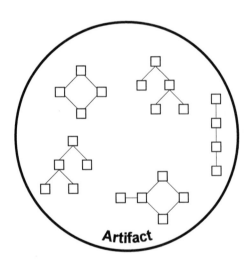

Fig. 5.1. The Artifact viewed as collective long-term memory for a group's tangible knowledge.

First, *long-term memory* is a permanent store. The artifact meets this criterion in a straightforward way. In ABC, information is stored in the graph structure and in the contents of nodes. This store is permanent, as much as any physical representation is permanent.

Second, long-term memory is frequently modeled as a semantic or associative network. An artifact built and maintained in ABC can be set up to resemble either type of graph structure. In both data models, nodes represent concepts and links represent relationships between concepts. Representing concepts as nodes with descriptive labels is straightforward. To represent relationships between concepts, ABC provides several options. Links can denote explicit semantic or associative relationships. Because all objects stored in ABC can carry attributes, attributes attached to links can define the link types required for a semantic network. For an associative network, no type scheme is needed. Consequently, ABC links can model either a semantic or an associative network. In its node/content relationship, ABC provides a second, hierarchical relationship between a superordinate concept and a cluster or a structure of subordinate concepts. Thus, the hierarchical properties of both memory models can be represented by either explicit tree structures or by the node/content relationship. A third type of relationship provided by ABC, called a hyperlink, can denote secondary semantic associations and connotations between concepts that are distant to one another in the main node/content hierarchy. Thus, a group's artifact can be implemented within an ABC-like data model so that it resembles either a semantic or an associative network and includes hierarchical chunking. Hence, an ABC artifact resembles the long-term memory components found in IPS models and architectures.

Third, information can be encoded, stored, and later retrieved from long-term memory. Because the artifact is dynamic, new concepts and new relationships can be added to it. In the ABC system, this means that new (labeled) nodes and new relationships — in the form of structural links, hyperlinks, or node/content relationships — can be added to the artifact. This is done by using special programs, called browsers, to work with the graph structure of the artifact or by using applications, such as text editors or drawing programs, to encode abstract concepts (represented by labeled nodes) as blocks of text, diagrams, or other forms. These tools, then, allow users to encode and store new concepts in the artifact. To retrieve information from this store, ABC users work with these same browsers to move from one concept or context to another, either by following hyperlinks (representing semantic associations) or by successively opening browsers and/or applications on the contents of nodes. Currently, ABC does not include a content search function, as would be expected in conventional models of long-term memory; however, such a function could be added. Thus, because information can be encoded and stored in an ABC-maintained artifact and later retrieved from it, the artifact satisfies this third requirement for a long-term memory.

Fourth, information retrieved from long-term memory is activated and processed in a working memory component. I claim this property here but delay justifying it until the next section.

Thus, the artifact, as maintained in an ABC-like collaboration system, is a large, permanent store that can be structured as a semantic or associative network, into which encoded information can be inserted and from which specific concepts can be retrieved or activated. Consequently, it can be viewed as a form of long-term memory for tangible knowledge.

Working   Memory

Working memory is the context in which information is activated and where conceptual processing takes place with regard to that information. In this section, I will focus on the component as a whole and defer discussion of the specific processes that operate within it until the next chapter.

As suggested in the preceding section, ABC browsers and applications can be viewed as a form of working memory with respect to information stored in the artifact. In IPS models, working memory has been described both as a separate cache that is loaded and unloaded with respect to long-term memory and as the currently activated parts of long-term memory. Researchers frequently shift between these two views as needed by context or for convenience. To simplify the discussion, I generally assume the cache version, although ABC tools can support either view. For example, a browser or application, as shown in Fig. 5.2, can be opened on any given subgraph or file of data that is part of the artifact. It presents the user with a view of the data and with functions for adding, deleting, and changing that information.   Thus, that segment of the artifact can be said to be *activated* in the browsers and applications because it can be consciously attended to and modified by users of ABC.

If one prefers, the relationship between long-term memory and working memory can be viewed as one of retrieving and storing, rather than activating.   That is, one can view the segment of the artifact presented by a browser or application as having been retrieved from the artifact.   The user can then perform various operations on this information using the functions provided by the browsers and applications.   When the user has finished his or her work on this segment, the (possibly modified) contents are returned to the artifact for permanent storage.

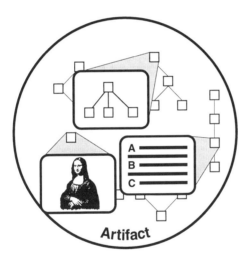

Fig. 5.2.   Browsers and applications viewed as working memory for a group's tangible knowledge.

Thus, ABC browsers and applications provide contexts where the contents of long-term memory can be consciously attended to and processed, and they can accommodate the two complementary views of working memory found in IPS models.   Consequently, we can regard them as a form of working memory for tangible knowledge.

Extended Memory

Extended memory is an external physical representation of information that functions as an extension of an individual's working memory. Often solving a problem, such as an arithmetic problem, is closely related to producing a sequence of intermediate products represented on paper. Or, when planning a document, many writers will jot down ideas on post-it notes during a brainstorming phase to relieve the burden on working memory and later organize those ideas into some sort of plan for the document. Both the post-it notes and the document plan can be viewed as forms of extended memory. ABC browsers and applications can also function as forms of extended memory for knowledge-construction tasks. For example, a graph browser can enable a user to represent and work with 40 or 50 concepts, as opposed to the four or five that can be dealt with in human working memory. Consequently, these tools function both as a form of collective working memory and as an extended memory for intangible knowledge.

Thus, we have identified constructs that are recognizable as approximations or extrapolations for the three memory systems found in cognitive models and architectures — long-term memory, working memory, and extended memory. All are used by collaborative groups to work with tangible information. It is, therefore, plausible to view these components as comprising a collective memory for tangible knowledge.

Intangible Knowledge

Not all of the information a group works with is tangible and, thus, stored in the artifact. A great deal remains *intangible,* carried in the heads of the individuals who comprise the group. In this section, I identify constructs that function as a memory system for tangible knowledge. It, too, consists of separate components for long-term, working, and extended memory, which, together, can be viewed as a collective memory for intangible knowledge.

Long-Term   Memory

Collaborative groups incorporate two forms of intangible knowledge: *private* and *shared*. Individuals are frequently added to a team because they have expertise or specialized knowledge not shared by all members of the group. Their *private knowledge* informs their actions and is used by them as they work on their parts of the artifact. Thus, it is a group resource, but it is not part of the group's collective knowledge because it is not immediately accessible to the group as a whole. In some cases, portions of this knowledge will be encoded into products, become part of the artifact, and thereby become part of the collective long-term memory for tangible knowledge. It is the residue that I want to look at here — knowledge that remains intangible but is no longer private.

Shared intangible knowledge is the intersection of the different collections of information stored in the individual long-term memories of the group members. It is developed and maintained by them in two ways. First, an individual member of the group who possesses private knowledge may transfer portions of his or her knowledge to the group as a whole. For example, one group member may brief the rest of the group on a technical topic. Second, the group as a whole may collectively build a segment of shared knowledge. For example, they may analyze a problem together and, in the course of their discussion, construct a solution. But, regardless of how it is derived, the body of intangible knowledge that is accessible to all members of a group can be regarded as a form of *collective long-term memory*.

Figure 5.3 illustrates the relationships among the artifact and both shared and private intangible knowledge. The artifact is shown in the center. It is surrounded by a layer of shared intangible knowledge, held in common by all members of the group; it is shown as the lightly shaded area in the figure. Around shared knowledge are the more extensive bodies of privately intangible knowledge, shown as darkly shaded areas, known to individual members of the group. This entire construct is dynamic, as new knowledge is constructed and new information is brought into the group, but it represents the total body of information that is available for use by the group.

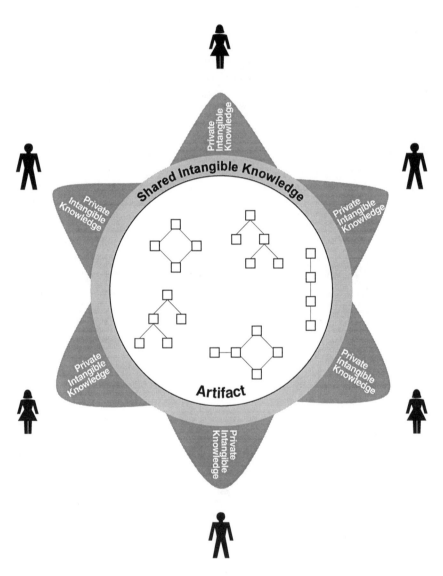

Fig. 5.3. Artifact surrounded by a group's shared and private intangible knowledge.

Working   Memory

The contexts in which intangible knowledge is activated and processed by the group form a *collective working memory* for this type of information. It is there that shared knowledge is built and there that it is refreshed and calibrated to maintain it within limits sufficient for it to function as a collective resource. The contexts in which this occurs are the social and intellectual situations in which a group collectively attends to particular segments of intangible knowledge. These situations range from chance discussions in the hallway to formal design reviews. Thus, we can regard activities such as these as a form of working memory for intangible knowledge, albeit a very abstract one.

The literal mechanism through which this abstract working memory is realized is the individual working memories of the individuals participating in the activity. From this perspective, the collective working memory is the union of the conventional working memories of the individuals taking part an event where shared intangible knowledge is developed. But that definition is not sufficient, because shared knowledge cannot be built in isolation. There must be communication, at the least, and, in most cases, knowledge must be collectively built and maintained by the group if it is to function as shared knowledge. Thus, a group's collective working memory for intangible knowledge consists of both the individual working memories of its members and the situations in which they activate their respective versions of shared knowledge and collectively attend to and/or modify it.

One of the most common and most important of these situations is the meeting. To see in what way it makes sense to regard a meeting as a form of working memory, let's look briefly at the way information is processed in those situations. Working memory provides the context for three basic kinds of processes: activation/retrieval, conceptual processing, and encoding/storage. Activation takes place in meetings as groups reconstruct shared knowledge. During discussions, groups frequently recall earlier discussions, decisions, or factual data to refresh and calibrate this knowledge or to incorporate it into new conceptual structures or decisions. The mechanisms for doing this are conventional. As topics arise during discussion, they key access functions, activating portions of members' respective long-term memories. If the topic is known to all, the activated structures are approximately the same, although not identical. However, if the issue is important, and hence memorable, the difference among versions are likely to be small. Thus, when I speak of shared knowledge, I am referring to an abstraction or an ideal that does not literally exist; what many groups do achieve, however, is an approximation that is sufficient to enable them to work coherently. Determining the degree of similarity among the different versions required for knowledge to be considered shared is a topic for future research.

As discussion within a meeting progresses, the conceptual structures currently activated in the working memories of individual participants changes. People add new ideas, argue against a previous point, forge new relationships between old ideas, and so on. Thus, a form of collective conceptual processing takes place.

Eventually, new information structures that evolve during meetings may be encoded and stored in the collective long-term memory — the union of individual long-term memories of the group members. This is more likely to be the case if the new construct is considered "important." New information that is shared in the moment but not regarded as important is likely to be forgotten by some or all of the participants and, hence, remains private relative to the group as a whole.

Thus, meetings provide a context in which shared knowledge can be activated, developed, and selectively encoded and stored in the collective long-term memory. Consequently, meetings can be viewed as a form of working memory where shared intangible knowledge is processed. Other forms of intellectual and social interaction serve a similar function; perhaps future research will identify them and examine them more closely.

Extended   Memory

In chapter 2, we saw that during meetings groups often use display devices such as a whiteboard to list points, to draw diagrams, or to write a sentence or a source code statement. Doing so makes visible the abstract concepts or structures being discussed. Consequently, such devices can help members remove extraneous differences in their respective understandings of what they are discussing, thereby contributing to the group's development of a common understanding of an idea. Consequently, ephemeral products such as these can be viewed as forms of *collective extended memory* with respect to a group's developing shared intangible knowledge.

Looking back at the chapter as a whole, I have identified two sets of constructs that function within collaborative groups as long-term, working, and extended memories. The first was concerned with the tangible knowledge a group works with and, hence, can be viewed as a collective memory for that type of knowledge. The second was concerned with intangible knowledge and, hence, can be viewed as a collective memory for intangible knowledge. Together, they constitute the collective memory for the group, in which all forms of information to which it has access as a collective resource are stored and processed.

Issues for Research

Several issues or questions for potential research arise from the preceding discussion.

- *What is the structure of a group's tangible long-term memory?*

Because it is tangible, the group's long-term memory, unlike human long-term memory, can be examined directly. What does it look like? What are its structural characteristics? Will groups build conceptual structures that resemble those thought to exist in human long-term memory (e.g., semantic or associative networks) or will they be structured differently? Will they build larger or smaller hierarchical components, perhaps influenced by the larger capacity of the collective working memory? To what extent will the total structure be kept internally consistent? How extensively will groups link "distant" concepts with one another through explicit semantic associations (such as hyperlinks)? Do differences in the degree of connectivity translate into greater or lesser coherence in the work of the group?

- *What is its pattern of growth and use?*

Because the collective long-term memory is dynamic, what will be its pattern of growth? Will it grow at a uniform rate for the duration of a project, grow by fits and starts, or grow quickly at first and then gradually decrease in rate? How will members of the group use it and participate in its development? Will material be copied and reused? Will individual members dwell in particular regions or will they range widely over the whole structure? Will work on one section lead to changes in another? Will internal consistency decrease with duration and/or size of the project? What problems will inconsistencies cause and how significant will they be?

- *How can we observe and characterize intangible long-term memory?*

Because the intangible long-term memory is maintained in individual minds, it is both distributed and unobservable. Are existing methodologies for studying it adequate? If not, which tools and methods can we develop so that we can "see" it and study it in close detail? Once we have the requisite tools, what will be its structure and form?

- *How is it developed, calibrated, and used?*

How will intangible knowledge grow and develop over the course of a project? Because it is more amorphous than its tangible counterpart, how will group members keep their respective versions consistent? How similar must these versions be to be considered shared? Will groups engage specific calibrating activities or will they allow portions of it to dissipate, resulting in a form of collective "forgetting"? How is intangible knowledge used by the group? How important is it?

- *What is the relationship between the tangible and intangible memory systems?*

How do the tangible and intangible memory systems relate to one another? Are there regular and observable patterns of development, where private knowledge becomes shared and, in turn, influences development of the artifact? Does the reverse process occur? What are the relative sizes of the two memory systems? Is this ratio consistent across groups, tasks, project sizes, durations, and so on? Do the two memories grow at similar rates? Does it matter?

- *What is the relationship between these systems and the strategic behavior of a group?*

How is the overall strategy of a group reflected in the growth and development of the two long-term memory systems? For example, a group that works by consensus might be expected to develop a large body of intangible knowledge in parallel with the tangible structure it is building, whereas a group that works largely by partitioning its work might develop a relatively smaller body of shared knowledge. Is this true? Can groups be trained or influenced in the ways they develop the collective memory and, if so, will it make a difference in their productivity?

Chapter 6

Collective Processing

A second major component included in the cognitive models and architectures discussed in chapter 4 is a processor. Consequently, we should expect a concept of collective intelligence to also include a *collective processor*. Just as conventional conceptual processing is done in close conjunction with the human memory system, so collective processing is closely related to the collective memory systems discussed in chapter 5.

The original Newell and Simon IPS model included a separate processor component as part of the basic architecture. In the Act* and Soar architectures, however, the processor has been replaced by a working memory component that functions as a cache and provides the context in which a set of processes operate on its contents. Thus, the emphasis has shifted from processor as object to the more abstract concept of a set of processes. I adopt this second perspective, focusing on the different types of processes groups use to get their work done, although for convenience I sometimes refer to these processes as the collective processor.

Before beginning the discussion, I want to point out an important difference in perspective between the IPS view of processor/processes and the collective processor/processes I am describing. Both Newell and Anderson were concerned with fundamental properties of human cognition and problem-solving behavior. Consequently, their models were implemented as autonomous computer systems intended to simulate human mental behavior and, thus, to function independent of the direct control of a human user. These systems are within the mainstream of artificial intelligence research.

By contrast, collective intelligence is situated within a different line of research, called *intelligence amplification*. This perspective emphasizes use of the computer to amplify or supplement human mental capabilities. Thus, a human user remains in control of the computer system. Furthermore, it is the human user, not the computer, that supplies the basic information processing operations required to carry out a task.

Just as the collective memory discussed in chapter 5 was comprised of separate subsystems for tangible and intangible forms of information, the collective processor also includes corresponding sets of processes used to develop these two types of information. A third type of process occurs in situations in which both types of information are simultaneously activated and developed by a group. The discussion that follows examines each of these three types of processes.

Processor for Tangible Knowledge

The conceptual processes used by members of a collaborative group to develop the artifact can be viewed as a *collective processor for tangible knowledge*. It includes operations carried out by individual members of the group working alone as well as the aggregate of their multiple independent activities. In this section, I will begin by focusing on a single individual working alone and then on multiple individuals working concurrently, but independently, within a group.

Individual Processor

Although the overall collaborative process includes situations in which the group works together in the same room at the same time, for many projects, the majority of time is spent in individual work in which members work independently at their respective desks and/or workstations. In chapter 2, I referred to this second mode of work as asynchronous collaboration. Consequently, a concept of collective intelligence must account for individual, asynchronous work on the collective enterprise as well as the group's more interrelated activities.

In recent cognitive models and architectures, working memory provides the context in which processes are applied to activated portions of long-term memory or to new input data. In chapter 5, I argued that the artifact, maintained as a hypermedia graph structure, can be viewed as a form of long-term memory for tangible information. Similarly, I showed that the various browsers and applications included in a collaboration support system can be viewed as forms of working memory. Consequently, the processes that groups use to develop tangible knowledge are closely related to the tools they have for working with the artifact.

In the context of a collaboration support system such as ABC, it is the individual user who supplies the basic mental processes responsible for representing new concepts within a browser, for perceiving and denoting new relationships between existing concepts, for building and modifying larger conceptual structures, for saving the results in the computer system's long-term storage facility, and for carrying out other information processing operations related to the artifact. The relationship between user as source of conceptual processes and the information to which those processes are applied is suggested in Fig. 6.1.

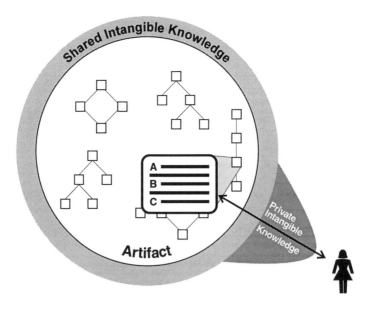

Fig. 6.1. Individual processor, working with an activated portion of the artifact, informed by both private and shared intangible knowledge.

In the figure, the tangible artifact is shown in the center. A part of it has been activated within a browser or application. The user, who, of course, is external to the computer system, is attending to that activated segment. As the user performs various operations on this information, he or she does so through a perspective shaped by his or her intangible knowledge. This includes knowledge shared with other members of the group, suggested by the lightly shaded area that surrounds the artifact, and knowledge held privately by the particular individual, such as specific technical expertise, indicated by the darker, uneven areas that surround the shared knowledge. Each individual member is associated with a different body of private intangible knowledge.

Work progresses through a sequence of changes made to the artifact. The basic unit of conceptual work involves a *cycle* of interaction between human user and supporting computer system. The primary function of this cycle is to keep the representation of information within the computer system consistent with the user's evolving mental state. Consequently, I call this basic cycle *computer-mediated cognition* (cmc) to emphasize the mediating effect the computer has on the user's thinking and to distinguish it from mental operations that take place without direct reference to representations maintained in such a system. The general form of this cycle is suggested in Fig. 6.2.

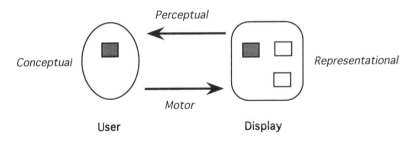

Fig. 6.2. Computer-mediated cognition cycles involve cycles of conceptual, motor, representational, and perceptual operations. The object of current attention is indicated by the shaded box.

The user interacts with the system through an ongoing sequence of conceptual, motor, representational, and perceptual operations.

The user's mental state may change as a result of some conceptual process independent of the current state of the computer system. When this occurs, he or she may elect to represent the change in the computer through a series of motor actions that control the system, such as moving a mouse or typing on a keyboard. When those actions are completed, the user perceives the change in the display, completing the cycle. In other cases, the conceptual process may be triggered by the representation, such as the user seeing that two concepts currently shown as independent of one another can be grouped to form a conceptual cluster. Again, a series of motor actions can be used to update the display in order to make it consistent with the user's new understanding of the data. Different forms of this basic computer-mediated cognition cycle are associated with different conceptual processes and produce different kinds of changes in the artifact.

To examine this concept of cmc cycles more closely, let's look at an example — the cycle used to add a new concept to the artifact — shown in Fig. 6.3. It describes the process by which a group member retrieves a concept from his or her (human) long-term memory and then represents it within the computer system. The cycle begins with a conventional memory access that produces a new conceptual object in the user's conventional working memory. I refer to this new concept as a "delta product" to indicate that it constitutes a change in the set of concepts currently available in the user's working memory for consideration; hence, it is shown in the figure as a "C" within a triangle or "delta" symbol. This process is shown in the first line of the figure.

Once the concept becomes available for attention, it is subjected to several tests, some of which may be unconscious or automatic. These tests — not necessarily in the order shown in the figure — determine whether or not the user regards the new concept as relevant to the current semantic context, whether or not it is worth representing, and whether or not it is already represented in the display. If any of these tests fails, this particular cmc cycle is terminated. If the concept passes these tests, a goal is generated by the user to represent it in the computer system.

To accomplish this goal, the user again accesses his or her long-term memory to retrieve a method by which to do so. It is comprised of knowledge of how the computer system works, consisting of one or more computer operations he or she knows how to use to create new objects in the display. Activating the method results in a sequence of motor actions performed by the user, interspersed with perceptual acts

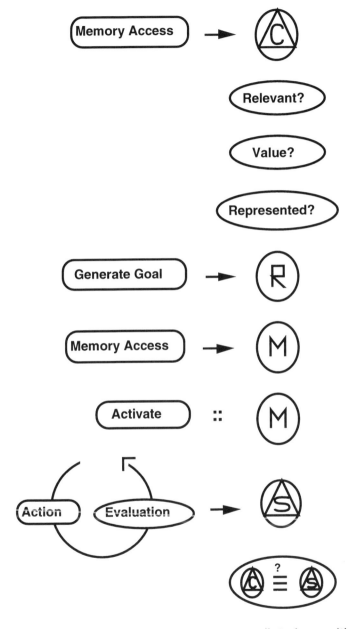

Fig. 6.3. Define_Concept computer-mediated cognition cycle.

to adjust those motor actions. For example, in an ABC network browser, the user points with the mouse to a position in the window, selects the *create node* option from a menu, types a word or phrase to denote the concept, and then types a carriage return to complete the operation. Successful completion of the method, thus, produces a new object within the computer system that corresponds to the conceptual object in the working memory of the user.

Once this new computer system object exists, it is subjected to one final test — to confirm that it is a satisfactory approximation of the original concept in the user's mind. If it is satisfactory, the cycle ends and another one begins; if it is not, an editing cycle is normally initiated (although the user could decide to live with the "unsatisfactory" representation).

Each such cycle constitutes one basic process within the overall set of processes that constitute the collective processor. Examples of other processes/cycles include relating two or more concepts to one another, adding a concept to a cluster or category of concepts, adding or deleting concepts to larger conceptual structures, and translating an abstract idea into prose or other form of expression. Complex knowledge-construction tasks require a number of different processes. For example, the task model for expository writing shown in Fig. 4.9 includes 21 different processes that occur in seven different cognitive modes. Different tasks require different sets of processes; however, at this very basic level, many tasks overlap. For example, many include one or more planning modes and, thus, share some of the same computer-mediated cognition processes used for abstract planning. However, tasks devolve into their respective idioms of expression, such as words, computer code, or technical diagrams, as abstract design is translated into concrete terms. Thus, we might speculate that the overall collaborative process will include tens, rather than hundreds, of basic cmc processes and a smaller number of more specialized cycles for particular tasks. A goal for future research is to identify the specific cmc cycles that comprise these collections.

In chapter 4, I suggested that under some circumstances we could consider the human user and supporting computer system a form of abstract, composite human–computer system. Computer-mediated cognition cycles can help us define this notion more precisely. Cmc cycles begin with a change in the conceptual structure(s) within a user's working memory, produced by some cognitive process. They also include a method through which the user updates the data in the computer system to make the representation consistent with his or her mental state. Thus, each cycle binds a cognitive process to a computer operation (or short sequence of operations that are used conventionally). It is this binding that allows us to think of the computer as an extension of the user's working memory, comparable to other forms of extended memory but more dynamic than most. Conversely, it lets us think of the user as supplying the essential information processing operations that change the artifact stored in the computer's storage system, which functions as a long-term memory for the group's tangible knowledge.

Ideally, the two subsystems should be closely tuned to one another. One way of doing this is to match the design of the computer's user interface and the functions it supports to the user's task model so that the user does not have to translate extensively between the two or to perform unproductive cognitive operations simply to control the computer. When this is the case and when the computer extends basic user capabilities, such as the number of concepts that can be attended to, it can be said to amplify the user's mental capabilities and, thus, his or her intelligence.

We can now define a general architecture for an intelligence amplifying human–computer system. It includes:

- a task model, expressed as a set of cognitive modes and the paths along which information flows from one mode to another
- a computer system that includes a corresponding set of interface modes and data transfer mechanisms between them
- a set of cmc cycles that bind task model to system model
- a human being who performs the task by using the system and engaging the cognitive modes identified in the task model

To be considered a true IA system, the task model must accurately reflect the user's mental habits. Thus, more than one system design may be required if different user populations perform the task differently, or, alternatively, the system must be adaptable to different users' strategies. This second approach may be preferable, since it would also allow the computer system to evolve along with users' behaviors as they become more proficient at both performing the task and using the computer system.

In summary, the processes that produce changes in the group's body of intangible knowledge can be described in terms of a set of computer-mediated cognition cycles. These cycles link a human user, who is the source of conceptual processes, with the components within a collaboration support system, such as ABC, that function as forms of working and long-term memory. This configuration is recognizable as an extrapolation of conventional cognitive models and architectures. Finally, when system model closely resembles users' task model, the system can provide a form of intelligence amplification.

Multiple   Independent   Processors

The processor described in the preceding section consisted of a single source of processing operations provided by a single human user of a collaboration support system. In this section, I discuss a collective processor that applies to the multiple individuals that comprise a collaborative group. The critical issue is to identify a concept of processor that can account for the independent activities of individual members yet also show how those various activities meld into a coherent whole that represents the overall behavior of the group. The discussion is limited to the processing of tangible knowledge, and I again assume that groups developing this form of information are working with a collaboration support system, such as ABC.

At least two notions of multiple processors are possible. First, the individual members could take turns functioning as the single processor described previously. One member would use the system at a time to work on the artifact. This would enable the group to utilize the specialized knowledge and expertise of its different members, and work could proceed 24 hours a day. However, a strategy or a system that restricted work on the artifact to only one member at a time would defeat many of the purposes for forming a group in the first place. At some point, there would simply not be enough working time to accomplish the task. Consequently, this mode of collaboration and the form of serial multiple processors it implies is not considered further.

A second notion of multiple processors — and the one that is examined in the remainder of this discussion — includes multiple individuals working on different parts of the artifact at the same time, permitting work to proceed in parallel. This view of multiple concurrent processors is illustrated in Fig. 6.4. The artifact is located in the center.   Multiple individuals in the group may work concurrently on different parts of it. The work of each is informed by the intangible knowledge he or she shares with other members, represented by the lightly shaded area, and by his or her private knowledge, represented by the darker shaded areas. To simplify the drawing, the working memory of each users is not shown but presumed as represented in Fig. 6.1.

This mode of work is made possible by systems, such as ABC, that support multiple concurrent users. Members of the group use browsers and applications to activate parts of the artifact, to make changes to those portions, and to save the modified segments. Within the terms of this discussion, each such tool constitutes a separate working memory, whereas the artifact represents the group long-term memory for tangible knowledge. Consequently, we can view each member using one of these tools as an independent conceptual processor, similar to the individual processor discussed in the preceding section. Thus, the multiple individuals who work with multiple instances of the system interface function as multiple conceptual processors.

While each configuration of a user, a working memory, and the group long-term memory resembles the IPS architectures discussed in chapter 4, what about the whole? What form of information processing system is it?

As already noted, this construct, from one perspective, is an extrapolation of the single processor model because it replicates processors and working memories around a core long-term memory, like the spokes of a wheel around a hub. But, unlike a wheel, coordinating the behaviors of these "spokes" is not straightforward. The multiple processors are bound to one another by the collaboration support system, but not rigidly so as the spokes of a wheel are bound to one another by the hub. Rather, they are more loosely coupled and, hence, it is harder to coordinate their activities and to see the coherence of their collective actions. Indeed, it is precisely this independent but coordinated form of behavior that permits a collaborative group to utilize the concurrent efforts of its members. Thus, there is a *fundamental* tension between the dependencies within a group that are ultimately responsible for giving their collective work coherence and the independence of their respective actions that enables work to proceed in parallel.

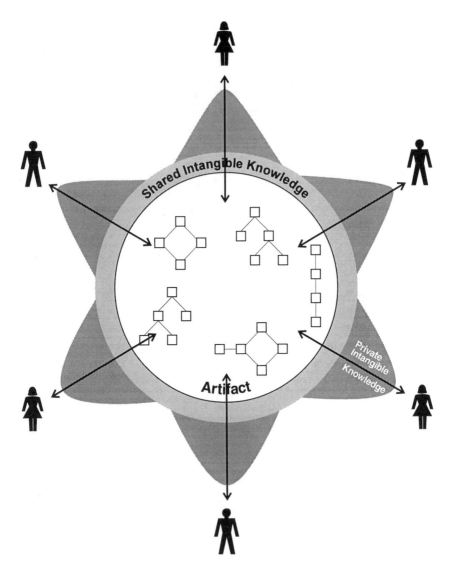

Fig. 6.4. Multiple independent processors form a collective processor for a group's tangible knowledge. Processing is informed by each individual's private and shared intangible knowledge.

To develop a concept of collective processor that includes multiple independent processors but is also coherent, we must extend the basic IPS architecture. This extension can be summarized in a simple proposition, which appears later. However, it will be easier to understand that proposition if we first place it in an historical context.

The original IPS perspective grew out of the predominant architecture for computer systems during the 1950s and 1960s. This so-called Von Neumann architecture, shown in Fig. 6.5, included a single processor, called the central processing unit (CPU), and a single storage component, called the memory. Data were loaded into the CPU from memory for processing and, after processing, results were returned to memory for storage. The CPU was also attached to input devices, such as card readers, and output devices, such as printers. This architecture grew out of the technologies available at the time and contemporary designers' understanding of control, the relationship between program and data, and other similar issues. It is ironic that these early machines were sometimes called "electronic brains" because they bore such little physical resemblance to that human organ.

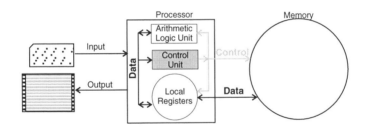

Fig. 6.5. Von Neumann computer architecture.

In spite of these differences, Newell and Simon built their IPS model of the human mind as an extended metaphor with respect to the Von Neumann computer. They abstracted away the physical structure of the machine to identify an underlying architecture that emphasized the storage, flow, and processing of information. That architecture provided a set of terms, components, and relationships in which to build a model of problem-solving behavior. If one compares the basic IPS model, shown in Fig. 4.1, with the Von Neumann architecture, shown in Fig. 6.5, the similarities are obvious. Looking at the human mind as analogous to a computer was, of course, a reductionist view.

But it allowed them to identify information processing as the *essential* activity and to use formal representations, such as computer programs, to describe basic conceptual operations. It also provided a vehicle, in the form of simulation systems, to test their models.

In the interval since the IPS model was first developed, several major new computer architectures have emerged. Those most relevant for this discussion includes multiple processors. They fall into two main families: tightly coupled and loosely coupled systems. Tightly coupled systems are more commonly referred to as parallel processor or multiprocessor systems. We can think of them as a single computer that includes multiple processors within it, each under the control of a central operating system. The processors communicate with one another by sharing a portion of computer memory and/or by exchanging messages through an internal bus. Other defining characteristics include the close proximity of the processors to one another — measured in inches up to several feet — and the system's vulnerability to component failure — when a processor or other component fails, the whole system usually fails (Mullender, 1989).

Loosely coupled systems are commonly referred to as distributed systems. They consist of a number of independent processors connected to one another by a communications network, such as a local area network (LAN); an example distributed system is shown in Fig. 6.6. Each processor is a complete computer, such as a workstation or personal computer. The system may also include specialized processors, such as file servers and high-speed compute servers (e.g., parallel processor computers) that provide services to the workstations or other processors. Each processor runs its own copy of an operating system. Thus, there is no overarching central control in a distributed system. Distributed systems also use shared memory and message passing as means of communication, but they differ from tightly coupled systems in that shared memory may be distributed across the separate workstations, rather than being centrally located, and messages are exchanged over the communications network, rather than over an internal bus. Thus, processors may be located at considerable distances from one another, measured in miles rather than feet. Distributed systems are also designed so that when one element fails, it does not cause the whole system to fail.

If we look at collaborative groups from an information processing perspective, they more closely resemble a loosely coupled

Workstations

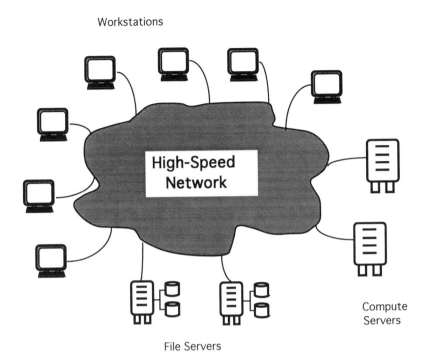

File Servers

Fig. 6.6. Loosely coupled distributed system.

distributed system than they do either the single processor Von Neumann architecture or a tightly coupled parallel processor architecture. Consequently, if we wish to develop IPS models of collaboration, a better starting point would be the distributed system architecture, rather than the Von Neumann architecture that served as the base for the original IPS models of individual cognition. This proposition can be summarized as follows:

IPS-Indv. : Von Neumann Arch. :: IPS-Collab. : Dist. Sys. Arch.

Translated into words, the proposition asserts that Information Processing Systems models of individual cognition and problem-solving behavior are to the Von Neumann architecture as Information

Processing Systems models of collaboration are to distributed systems architectures.

In effect, this proposition asserts a metaphoric relationship between collaborative groups and distributed systems. We should not confuse metaphor for model, but a metaphor can help us consider collaboration in several ways. We are led to a set of suggestions in the form of a set of issues important in one domain that may be important in the other. A metaphor can also provide terms, relationships, and visual images that can be tried and then retained or discarded, as warranted. And it can provide a framework that can serve as a starting point for the structure of a theory of collaboration. In the remainder of this section, I want to explore this metaphor more closely by looking at several characteristics of distributed systems and then at similar properties for collaborative groups. (For a more thorough discussion of these and other issues associated with distributed systems, see Coulouris & Dollimore, 1988; Mullender, 1989; Mullender, 1993).

Parallel Processing. The core concept of a distributed system is parallel, independent processing. Because each processor is a complete, separate computer running its own copy of an operating system, each is capable of operating autonomously. From a task perspective, this leads to two concepts of asynchronous parallel processing. First, users may work completely independent of one another if each is working on a separate task unrelated to the task any other user is working on. In this case, the only thing logically joining these users and their workstations may be the underlying file system that is part of the computing infrastructure. A second type of parallel processing can occur when two or more workstations are used to work on the same task. This can occur for tasks that can be decomposed into independent operations that can be carried out separately and whose independent results can be merged to form the solution or an intermediate result. Matrix multiplication and decryption are examples of such tasks. The critical issue for this second type of task is finding a valid and effective decomposition.

Parallel processing is also a fundamental part of most forms of collaboration. As I have emphasized, intellectual groups are usually formed because the task is too large and/or too complex to be done by a single individual. Therefore, to meet time requirements, group members must work in parallel. This mode of work may sometimes resemble the separate task form of parallelism described previously — for example, when team members separately write different sections of a document — although independent work must always be synthesized in a collaborative group if the work is to be coherent. On the other hand, when critical expertise is held by only a few members of a group, it is often necessary to form teams whose members have complementary skills and have them work together closely on the same task. This second mode of work resembles the same task form of parallelism described previously. Thus, finding an effective task decomposition is as critical for collaborative groups as it is for distributed systems. Key issues for research, then, include how groups divide their work into independent steps that can be carried out concurrently, how large these steps should be, and what constitutes a good versus a bad decomposition.

Communication. If a system is composed of multiple independent processors, those processors must be able to communicate with one another if the system is to function coherently. Processors must be able to issue and respond to messages, update a common store of information, and report their current status. In a distributed system, these interactions are carried out using a communications network to which each processor is linked. However, for effective communication, processors must also understand various conventions and protocols. Some of these are rather mechanical, such as formatting information in a particular way for transporting from one location to another. Others are more social — for example, agreeing to acknowledge receiving a message before carrying out the requested operation, analogous to rules of courtesy in carrying on a conversation. Still others have to do with the integrity of the data, such as guaranteeing that the requested operation will produce a particular change or else no change at all to stored data. Thus, communication covers not just the channels over which messages are exchanged, but a complex, multilevel structure of conventions, agreements, and guarantees.

Communication is obviously an extremely important part of the collaborative process. Information flows between members of a group over the same communications network used by distributed systems in the form of e-mail, ftp files, and other forms of electronic communication. But it also flows through a more complex web of human relationships, both formal and informal. This information occurs in a variety of forms, ranging from memos and meetings to chance encounters in the hallway and back-door calls to a friend to get something done. Each of these interactions can serve useful, even necessary functions, and each has its own social conventions.

Important issues for research are documenting the different types of communication that occur within groups — along with the conventions guarantees, etc., that are part of communication — and understanding how they contribute, individually and as a whole, to the work of the group, especially its efforts to construct a coherent artifact.

Fault Tolerance.   No physical system is perfect or will remain operational indefinitely.   What designers aim for are systems that will fail gracefully.   That is, when a component of the system fails, it may inconvenience some users or degrade performance, but it will not result in catastrophic failure of the entire system.   A related issue is component replacement.   One must be able to replace a defective component or upgrade a component to take advantage of technical advances without affecting the entire system or its overall design.

Groups must also be fault tolerant.   They must be able to survive the failure of an individual or team to accomplish an assigned task. Such failures must first be recognized, through monitoring work, checking milestones, or other means, and the group must have the means to correct the problem or replace the component responsible. Similar problems occur when a member leaves the group or becomes incapacitated and must be replaced.   A somewhat different problem occurs when a group discovers that it needs expertise not currently available within its membership and must replace a member or add a new member to provide it.   This last situation resembles component upgrade.   Documenting the mechanisms groups use to provide different forms of fault tolerance is an area for further research.

Transparency.   Transparency in a distributed system is concealing aspects of the system from a user so that it appears to be a whole rather than a collection of separate parts.   There are a number of different types of transparency (Coulouris & Dollimore, 1988). Access transparency, for example, enables a user to work with files stored on a workstation's local disk and those stored in a remote file server in the same way.   Location transparency enables files and computer services to be accessed without having to know where they are physically maintained.   Replication transparency hides the fact that multiple copies of an object may exist at different locations to improve access and reliability.   Failure transparency allows users to continue work despite the failure of a particular component, such as a server. Scaling transparency allows the system to expand without having to change its structure or the way users work with it.

The analog of transparency in a collaborative group is concealing aspects of the group from an outside observer so that it appears to be a coherent whole rather than a collection of individuals. Perhaps the most important form of transparency relates to the artifact — that it appears to be the work of a single good mind, rather than a number of separate hands, in its integrity and consistency. We might call this property *artifact transparency* to indicate that the work appears seamless. Analogs for other types of transparency can also be recognized in groups. Expertise critical to a group may be replicated in several members to insure its availability and to provide continuity should a key member leave the group. Similarly, when someone outside a group requests something of the group, such as information or an action, he or she should not have to know which individual in the group will answer the request. Perhaps the hardest form of transparency to meet in a group is scale. When a group grows in size, the change is often visible both inside and outside the group. If people are added to a project because the group is behind schedule, the result can often be a slow down rather than an increase in productivity, caused by the overhead in bringing new members up to speed and the additional complexity in communication and coordination that results from a larger group (Brooks, 1975). Inside the group, members may feel that the character of the group has changed, particularly if it goes from being a small, close-knit team to a significantly larger organization with more formal structure and lines of authority. Thus, techniques and designs that permit scaling and other forms of transparency in distributed systems may also be useful in organizing and supporting groups to provide similar properties, such as those discussed here. Testing that possibility could provide an extensive research agenda

Shared State. Shared state refers to a body of current information two or more processors maintain in their local environments about one another or about the system as a whole. The individual components of a distributed system must maintain a certain minimal amount of information about other components in order to recover from component failures. Otherwise, part of the overall state of the system would be lost when a given workstation or server failed (Mullender, 1989). At the other extreme, determining in any very detailed way the overall state of a distributed system is difficult, if not impossible (Birman, 1989). What one would like to have is a global view of a distributed system, such as that of an imaginary ideal observer who could look down on the system as a whole and see into the operations of its individual processors and the communications network. No such ideal observer can exist in fact. We are forced to observe distributed systems from within, by having processors report their individual states to a monitor processor. However, between the time events are reported by individual processors and the time those messages arrive at the monitor process and it updates its representation of the global state, a host of additional changes can take place in the system. Thus, because of interdependencies and the time required to exchange messages, one can never be sure that the current state of the system as recorded in the monitor reflects the state as it truly exists at that moment.

Maintaining shared state is also both important and difficult in collaborative groups. The body of shared intangible knowledge a group develops can be viewed as a form of shared state, comparable to the type necessary for fault tolerance discussed previously. On the other hand, it is also difficult to ascertain the overall state of a collaborative group in a detailed way, just as it is for distributed systems. We, too, would like to have an imaginary ideal observer who could look down on a group as a whole and make sense at any given moment of all of the various actions taking place within it. But no such observer is possible. Instead, we have to rely on milestones, reporting procedures, and the intelligence and good will of group members to achieve coherence in collaborative work. I look at this issue in more detail in chapters 7 and 8 in discussing collective strategy and collective awareness, respectively.

Parallel processing, communication, fault tolerance, transparency, and shared state are all important properties of distributed systems that have analogs in collaborative groups. There are numerous other correspondences. For example, both distributed systems and collaborative groups must *synchronize* the actions of their various components/members. Both *replicate* important information in multiple locations/in multiple members. Some include *heterogeneous* components — distributed systems, to permit different types of hardware and software to be used together; groups, to have access to different skills and bodies of expert knowledge held by different individuals. And both carry out basic operations on tangible data in series of discrete operations — *transactions*, in distributed systems; computer-mediated processing cycles, in collaborative groups.

By recognizing this rich metaphoric relationship between distributed systems and collaborative groups, we are led to a number of suggestions for research and for building a theory of collective intelligence. One form these suggestions take is a set of issues that are important in one domain that may be important in the other. They are not the only issues that should be addressed or even necessarily the most important ones, but they are a ready list that should be checked out. A second type of suggestion is a general framework in which to build theory. This framework is the architecture of distributed systems. It may not be the one that, ultimately, is most appropriate for a theory of collaboration, but it can serve as a starting point. As we study groups and build knowledge about parts of the collaborative process, we can replace the analogous component in the framework with the real thing. By the time we are through, it may be that no piece of the original is left, like replacing the boards of a house one by one. But we will have been guided in our enterprise by a shape of the whole.

To get a more specific sense of what an IPS architecture for collaboration based on a distributed systems architecture might be like, consider a collaborative group using the ABC system. ABC, itself, is a distributed system, as is the component that stores and manages the artifact, the Distributed Graph Storage System (DGS). Although logically central, the DGS is implemented as a set of independent processes that run in parallel on multiple workstations and can store data in separate file systems. The browsers and applications that users work with run as independent processes on multiple workstations. Because they can all access the artifact, these programs, or processes, can be said to communicate with one another using a form of shared memory. In ABC's computer-conferencing facility, they also communicate with one another through a form of message passing.

What is missing from this picture are the members of the group. In the preceding section, I described a concept of intelligence amplification in which human user and computer system are so closely bound to one another that the computer can be viewed as an extension of the user's mental apparatus. It is a straightforward extrapolation to consider the multiple members of the group as being similarly bound to a system, such as ABC. Each member of the group can then be viewed as an individual processor (or source of processing operations) that works on the contents of the collective long-term memory, the artifact. This is done by activating portions of the artifact in individual working memory components — the browsers and applications. It is this composite human–computer system, in which a distributed group of individuals work with one another using a distributed collaboration support system, that we need to describe in order to formulate an Information Processing System model or architecture for collaboration. And, I suggest, it is this composite human–computer system that may achieve collective intelligence.

In this section, I suggested that to build an architecture for collective intelligence, analogous to a cognitive architecture for individual intelligence, we should begin by considering the architecture of loosely coupled distributed systems. In doing so, we inherit a richly suggestive set of issues that can guide and inform research. We also inherit a general framework that can serve as a starting point for such a theory.

To this point, the discussion has focused on the processing and development of tangible knowledge. To understand the overall collaborative process, we must also understand how groups develop intangible knowledge and how that knowledge affects their more tangible work.

Processor for Intangible Knowledge

In chapter 5, I suggested that the intangible knowledge shared by the various members of a collaborative group can be considered a second form of long-term memory. I also suggested that the various situations in which members of a group discuss and develop shared knowledge can be viewed as forms of a collective working memory. Meetings served as the representative example of these situations. In this section, I consider a form of processor that provides the developmental component for creating and using shared intangible knowledge. Because intangible knowledge is stored in the heads of the members and because the situations in which it is developed all involve human participants, we should look to this same group of collaborators as the source of the various processes used to build and maintain intangible knowledge.

Developing the collective intangible long-term memory is, by definition, an activity that involves multiple human beings. It is theoretically possible that one individual could generate the entire construct, convey it to the group, and the others merely remember it; however, this behavior is neither desirable nor possible in actuality. Thus, I assume that developing the shared intangible long-term memory is a collaborative process in which all members of the group participate.

In situations, such as meetings, in which intangible knowledge is developed, individuals see events from their own individual points of view. However, the situation, itself, inevitably exerts a strong mediating effect on individual cognitive and conceptual processes. That is, the thinking of each individual is inevitably influenced by the thinking of the other members taking part in the activity, even if it is only to disagree. I refer to this situated form of thinking as *group-mediated cognition* (gmc).

Group-mediated cognition takes place within basic cycles of interaction between the individual and the group. Some gmc processes are (almost) entirely intellectual. Others are primarily social. But many, perhaps most, include both conceptual and social dimensions. For example, an idea voiced by a member of the group is evaluated not just on its intellectual merits but also in accord with the listener's assessment of the person voicing it. This merger of intellectual and social processes is one of two fundamental properties of group-mediated cognition.

A second fundamental property is the tension between the individual and the group. More precisely, it is the tension between the conceptual structure that is held in common and, thus, is said to be shared and the slightly different versions of that structure that exist in the individual working memories of the participants. This tension provides both the energy and the developmental operations that drive this form of collective processing. For example, all participants may share the same core structure, but in some minds parts of that core structure may be linked to additional concepts through private associations. If an individual views these associations as inconsequential or idiosyncratic, he or she is likely to remain silent about them. But, if the individual views them as relevant and feels comfortable speaking up, he or she may describe this "new idea" to the group. The other members hear these comments, apply them to the structures in their respective working memories, and thereby change those structures. When this occurs, shared intangible knowledge is extended.

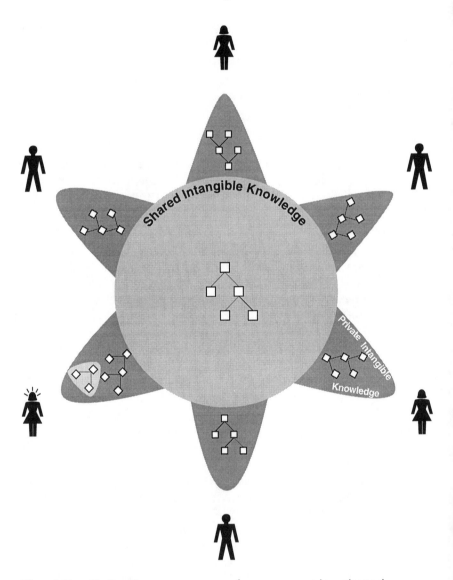

Fig. 6.7. Collective processor for a group's shared intangible knowledge. All participants share the same basic structure of ideas, but one has extended the structure.

The situation just described is portrayed in Fig. 6.7. It provides a context for an *add_concept_to_discussion* gmc cycle that will serve as the example process that illustrates the concept. A small conceptual structure — such as an earlier decision or a portion of a design — has been brought into the discussion and, thus, activated within the collective working memory of the group. It is shown in the middle of the figure. The individual working memories of six participants are shown at the corners. Although all six versions of the core structure are essentially the same, their orientations are different, suggesting minor differences in perspective. The individual shown at the lower left has made a more basic change in her version by realizing that one of the concepts is related to two additional ideas. She is about to describe the new insight to the group.

Fig. 6.8 shows a sequence of operations for the add_concept_to_discussion process. The cycle begins at the point where the individual has just constructed the extension to a concept included in the core structure. This realization is represented as a change in that individual's conceptual structure and, hence, is identified in the first line of the figure as a "delta concept" operation.

Once the individual is aware of the change, he or she subjects it to two different kinds of tests — one conceptual, the other social. Conceptual tests address the content of the change. Is the change new, or has it already been mentioned by someone else? Is it relevant to the discussion? If so, is it worth mentioning? A parallel set of tests assess social aspects of the situation. Do I feel comfortable speaking? How will the group react to what I may say? Are the potential benefits or losses from speaking worth the risk with regard to my position in the group? These tests are not necessarily performed in the order shown in the figure, and the individual may not be consciously aware that he or she is applying them. One should also note that the two dimensions, conceptual and social, are interdependent, as, for example, seen in the (potential) speaker's assessment of how the group is likely to react to the new concept. If the idea passes these tests, then the individual is likely to make a decision to speak up and describe the new idea to the group.

Once that decision is made, the person will probably spend a few moments formulating a statement. Some people rehearse the actual sentences they plan to say. Others mentally prepare a list of points, but at an abstract level. Still others encode as they speak with little or no prior planning. Regardless of the particular tactic, various planning and/or encoding processes will be used at one or more times to transform the initial change in the core structure into the statement or communication that is subsequently made to the group.

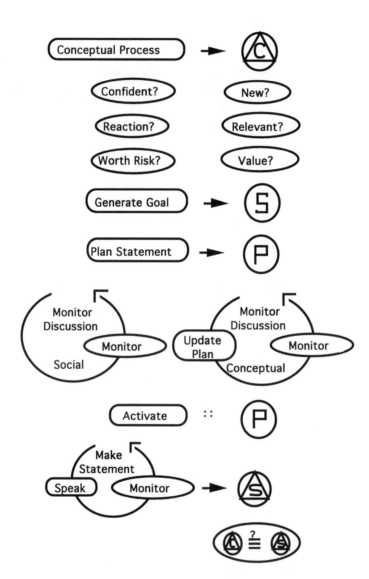

Fig. 6.8.   Add_concept_to_discussion group-mediated cognition cycle.

In most meeting situations, people cannot just speak when they are ready.   Rather, they must follow some social or organizational protocol to gain the floor.   At one extreme, this can be a formal process of asking the chairperson for recognition. More often, the process is more informal — the individual waits until the current speaker finishes and then speaks up or, if he or she is beaten to the punch by someone else, iterates the monitoring procedure.

During these periods of waiting for a chance to talk, the individual must also monitor the contents of the discussion. Monitoring operations include comprehending the statements made by others, mapping their ideas onto his or her own ideas, and, if need be, updating both the idea the person intends to talk about as well as the plan for the statement itself.

Thus, at least three complex, even contradictory, processes operate during the interval of time between awareness by the individual that he or she has something to say and actually saying it: planning or encoding the statement, monitoring the social or organizational situation to gain the floor, and monitoring the intellectual content of the discussion and reconciling it against the planned statement. Although these processes seem to operate in parallel, further work is needed to determine whether this is so or whether people switch back and forth among them.

Once the floor is gained and the person begins speaking, he or she is likely to monitor the responses of the other members of the group to see if they are following, if they signal agreement or disagreement, and so on. Results of this monitoring procedure can lead to on-the-spot modifications to the statement — for example, the speaker might go into further detail on a point thought not to be understood by the group, or the statement might be cut short if the person feels it is producing a negative reaction.

Once the statement is made, the group may coalesce around it. This may occur spontaneously, or consensus may gradually emerge as influential members voice their agreement. On the other hand, the idea may be ignored as the next speaker begins his or her statement. And, of course, other intermediate scenarios between overt acceptance and rejection are possible. Regardless of the particular scenario, the speaker is likely to monitor these reactions to determine, first, whether or not the other members accurately understood the idea and, second, whether or not the idea is being accepted. If it is accepted, it will probably be remembered and, thus, become part of the core conceptual structure that is shared by the group. In this case, this particular gmc cycle is complete. If not, the individual must decide whether to pursue the point, to come at it from another angle, or to say nothing and leave the idea to its own fate.

Acceptance by the group, however, is not a requirement for a concept to be remembered by the group. For example, if the group is divided between two points of view on a topic, the discussion may become a debate. If the discussion is memorable and the members share a common understanding of what was said, then the concept can become part of the group's shared knowledge, regardless of whether individuals agree or disagree with it.

Thus, the add_concept cycle is a basic process by which private intangible knowledge becomes shared intangible knowledge. However, it is not the only such operation. Other gmc cycles include establishing new relationships between shared concepts; replacing a concept or a portion of shared knowledge with an alternative concept or structure, through argument or persuasion; translating an abstract concept into some more specific form of expression, such as words, code, or diagrams. Additional research is needed to refine and extend this list of gmc cycles and to articulate the precise sequences of operations that comprise them.

Ephemeral products play a particularly important role in the development of shared intangible knowledge. For knowledge to be shared, the different versions stored in the respective long-term memories of the members must all be approximately the same, but they will not, of course, be identical. Indeed, the developmental process depends upon such variations. The goal, then, is to minimize extraneous differences without constraining the diversity of opinion and individual intellectual contributions that are crucial to the collaborative process.

Ephemeral products help groups achieve a balance between similarity and diversity. For example, during a discussion, when an individual sketches an architecture for a computer system on a whiteboard, the other members can literally see that person's ideas and perspective. Thus, they all share the same image. If that structure is accepted by the group, it is likely to be encoded and retained in their respective long-term memories and, thereby, becomes part of the collective long-term memory. Often, however, the diagram evolves during discussion. For example, someone adds a new component, another changes it to make it interface with another part of the structure, a third re-draws the diagram to simplify it. When this happens, the whiteboard becomes a common field where the group shares the same structure of ideas and where all members can make their individual contributions. Thus, ephemeral products tend to remove the noise of accidental variations inherent in separate versions of shared knowledge while admitting the free exchange of different points of view.

Finally, I note the similarity between group-mediated cognition cycles and the computer-mediated cognition cycles discussed earlier. Some individual cycles are similar, although not identical, in form, as can be seen by comparing Fig. 6.3 and Fig. 6.8. As a group, the two sets of cycles are similar in function, as the active elements used to develop their respective types of information. A topic for research is to explore similarities and differences between these two types of mediated cognition cycles.

In summary, group-mediated cognition cycles can be viewed as context-sensitive processes that enable a group to develop and maintain a body of shared intangible knowledge. During discussions and other situations where this type of knowledge is developed, participants activate portions of their respective long-term memories, each of which includes both knowledge held in common with the rest of the group as well as private knowledge known only to that individual or to a subset of the group. Once activated, the common structure of ideas evolves through discussion or other similar forms of interaction. The processes responsible for this evolution exhibit two fundamental properties. First, most gmc processes merge social and conceptual operations. Second, the tension between private and shared knowledge is essential for development to take place, although ephemeral products mitigate this tension and mollify extraneous differences. Eventually, a modified version of the shared conceptual structure is encoded and stored in individual long-term memories and, thus, merges back into the collective long-term memory of the group. Thus, group-mediated cognition cycles constitute the basic set of operations that comprise a collective processor for shared intangible knowledge.

Hybrid Processor

In this section, I look briefly at a third type of processor, which I refer to as the *lxn hybrid processor*. It includes multiple processors operating with respect to a single working memory in which a portion of the artifact has been activated. In these situations, intangible knowledge is also activated and developed.

One form this processor can take is for two or more members of a group to gather around a single workstation to confer about an issue that involves the artifact. They would initiate a browser or application and then either take turns operating the workstation or one member would operate the workstation for the group.

We have observed this behavior in our lab among members of a programming team (Kupstas, 1993). A group of five programmers worked together on a common project in the same room at the same time for some 4 to 6 hours a day over a 3-month period. The workstations were arranged in a "U" with the team members sitting on the inside of the "U". Most of their time was spent in individual work; however, they also interacted with one another at fairly regular intervals, normally for brief periods of time. A common form of interaction was for one member to call over to another to ask for specific information, such as the name of a file or the status of a task. They also frequently "wheeled over" to one another's workstations (they used office-style chairs with wheels) to discuss information displayed on a screen. During episodes of this second type, one member would normally control the system and the other member(s) would view the screen, talk about its contents, sometimes pointing to specific information. Most of the time, the purpose of these encounters was to transfer information — it was easier for the person with the knowledge to show the person than to tell him or her what was wanted. Occasionally, however, they would add new information to the artifact by editing a paragraph or a function, by changing a diagram, and so on. This latter behavior can be considered a hybrid form of processing because the members were both building shared intangible knowledge as well as activating and working with the tangible artifact.

A second form of hybrid processing takes place in computer-supported conferences. Instead of several group members gathering around the same workstation, conferencing systems — several of which were described in chapter 3 — permit participants to work from their respective workstations. Figure 6.9 shows the logical organization of this type of interaction.

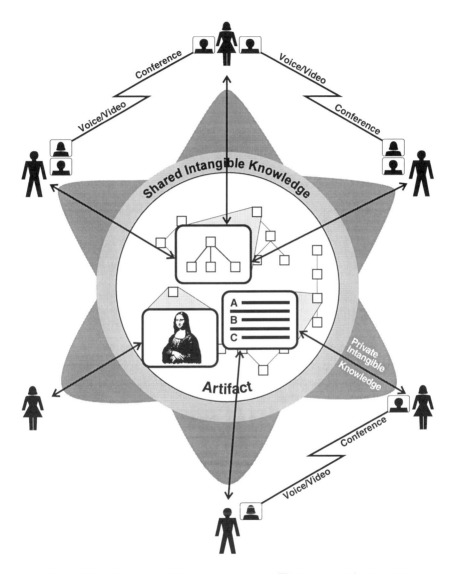

Fig. 6.9. 1xn hybrid processor. Three members are engaged in a computer conference while two others work independently. Conference participants can also see and talk with one another through voice/video communications

When a group works with a conferencing system, one widely used architecture includes a single copy of a browser or application program that runs on a single workstation but selectively takes input from the other participants' workstations and "broadcasts" the program's output to all of their workstations. Although at any one time only a single user normally has control of the system and thus provides input to the program, over a working session, control is normally shared among the participants by passing a symbol, called a *token*, from one user to another — whoever has the token becomes the active user. Some conferencing systems include supplemental voice and video channels that allow participants to talk with and see one another as they work on the artifact. The challenge facing collaboration system builders is to make this distributed form of conferencing as easy and as natural as that of groups working in the same room, such as the group described by Kupstas.

This type of computer conferencing combines aspects of both the processor for tangible knowledge, described in Fig. 6.4, and the processor for intangible knowledge, described in Fig. 6.7. It includes operations that directly affect the artifact and, hence, the group's tangible knowledge. Because participants can also see and talk with one another and, thus, discuss what they are doing, it also resembles the processor for intangible knowledge.

However, the configuration also differs from both of these processors. Because discussion can refer directly to the artifact, it is likely to be more grounded and more objective than conventional discussions that take place in meetings and other situations where the artifact is less accessible. Second, because a computer conference incorporates the private knowledge of its multiple participants as well as the knowledge they share, work on the artifact that takes place in a computer conference is informed by a larger body of intangible knowledge than work performed by an individual member working alone. A topic for research is to document these differences more precisely and under a variety of conditions.

In this section, we have considered two forms of hybrid processor — an assembled form and a distributed form. In the assembled form, a group worked together in the same room at the same time. Interaction was completely natural, and the members moved from individual to collective work easily and instinctively. In the distributed form, behavior is mediated by the technology. Consequently, the ease and naturalness with which members shift between individual and collective work is determined to a great extent by the design of the support system(s) and current limitations of that technology. These two modes of behavior represent points along a spectrum. However, they are sufficiently alike that we can consider them to be a third type of collective processor. A goal for research is to develop detailed descriptions of this mode of collaboration, including *conference-mediated cognition* cycles that function as the basic processes that operate within it.

In this chapter, I have described three types of processors found in collaborative groups. First, the multiple independent processor supports the individual work of group members as they work on various parts of the artifact. Whereas each member working alone functions as an individual processor — in effect, a single user working with an intelligence amplification system — together they form a multiple processor that resembles a loosely coupled distributed system. Second, the processor for intangible knowledge combines both conceptual and social processes. It, too, includes multiple processors oriented around of common body of shared knowledge. However, because that knowledge is stored in the respective long-term memories of the different members of the group, to say that it is "shared" is only an approximation. Although differences in members' recollections can cause problems, they are also what makes it possible for the group to develop shared knowledge. Third, the hybrid processor combines aspects of both the tangible and intangible processors. Together, these three processors can be viewed as the *collective processor* component of a collective intelligence.

When we attempt to build actual IPS models of collaboration, those models should be based on the architecture of a loosely-coupled distributed system, rather than the Von Neumann architecture that provided the basis for earlier IPS models of individual intelligence.

Issues for Research

Several issues for research are suggested by the preceding discussion. They include the following:

- *Identify a target set of collaborative tasks to be examined.*

If, as a research community, we are to build on one another's work, it is important that we be clear about the domain of study for any given research project. Consequently, as a project begins a study, it should make clear the type of activity and the type of group being examined so that others can know where within the overall space of collaborative behavior the observations and results fit.

- *Identify activities that occur in the target set.*

Once the particular type of collaboration has been determined, as suggested by the preceding issue, a second step is to identify the large-grain behaviors that occur within the group(s) being studied. For individual work, studies of this sort could identify the various cognitive modes used by members to carry out particular tasks. For group work, this would involve identifying the habitual activities the group engages in. Can we identify a "vocabulary" of such behaviors? Is there overlap across groups and tasks with respect to these behaviors?

- *Identify specific mediated cognition cycles.*

The next step is to identify the individual processes that occur within these larger activities. When an activity is carried out by someone working with a computer, basic processes can be associated with a set of computer-mediated cognition cycles. When an activity is carried out in a group situation, such as a meeting, they can be associated with a set of group-mediated cognition cycles. And, when an activity is carried out in a computer-based conference, they can be associated with a set of conference-mediated cognition cycles. Detailed descriptions of the cycles that operate in each of these three processors will provide a fine-grained view of both individual and group behaviors for particular forms of collaboration.

- *Construct an IPS architecture for collective intelligence.*

We need an architecture for collective intelligence analogous to the IPS architecture for individual intelligence. It must include both an overarching framework as well as a set of basic constructs that can be used as building blocks with which to build specific models of collaborations. One promising approach would be to develop this architecture as an analog to the loosely coupled distributed system architecture that includes multiple independent processors coordinated through shared memory and message passing protocols. This will require a substantial base of prior knowledge about collaborative behavior. But we have a major asset in the examples of Newell, Simon, and Anderson and their development of the original IPS models/architectures. Although the process will not necessarily be the same, we can learn from their experience and methods.

Chapter 7

Collective Strategy

In chapter 6, three types of processes were discussed that are used by collaborative groups to develop the contents of the collective long-term memory. These included computer-, group-, and conference-mediated cognition cycles, comprised of short sequences of social and/or cognitive actions interspersed with evaluative tests. The emphasis there was on identifying the underlying architectural form of these processes and to illustrate that form through example cycles. Each process or type of process was treated separately, and no attempt was made to consider the relationship of one process to another as they occur over time and within the flow of a group's work. In this chapter, I look at patterns among processes that result from and reveal the strategies groups use to accomplish their goals.

This distinction between the structural form of processes and their situated occurrences is similar to the distinction computational linguists make between types and tokens in language. *Types* refers to the words in a vocabulary for a given language; *tokens* refers to individual occurrences of those words in sentences or other expression generated by a speaker or writer of that language (Herdan, 1960). For example, the unique configuration of letters — *c a t* — denotes the word type, *cat*, whereas its occurrence in the sentence, "The *cat* scared the dog" represents a token or instance of the type. Thus, there may be many tokens of a given word type in a particular sample of text.

The set of collaborative behaviors known to a group is analogous to the set of words that constitute a vocabulary; an instance of one of those behaviors within a group's flow of work is analogous to a token within a particular expression. Just as words are not used randomly to communicate an idea, but in coherent expressions, so behaviors within a group do not occur randomly, but are composed into coherent, purposeful sequences.

156

The issue, then, is to identify the patterns of behavior that occur in collaborative groups that give coherence to the actions of their individual members and, in turn, to the artifacts they build.[1] How are coherent sequences of actions generated? How can we, as researchers, recognize them and understand their effects? These questions go to the heart of collaborative work. We form intellectual groups to increase the pool of effort available for tasks too large to be done by one person and to increase the pool of expertise for tasks that transcend the knowledge and skills of any single individual. Consequently, groups must be able to decompose tasks into separate subtasks that can be done by individual members or teams; but groups must also assemble the separate pieces of work that result from decomposition into a coherent whole. If the synthesized artifact is to be coherent, the group must, in general, work coherently. Good work does not happen by accident, at least not for large projects. To argue otherwise would be to argue that we should be able to routinely throw paint at a canvas and expect to produce artistically valid paintings — theoretically possible, but not very likely.

Coherence is not, of course, a black or white matter. All groups exhibit at least a minimum degree of coherence; otherwise, they would not be regarded as a group in the first place. Rather, it is a matter of degree. It is reasonable to expect that as groups exhibit more or less coherence, they also work more or less efficiently and produce higher or lower quality products. If this hypothesis turns out to be true, it would indicate that technology and training that increase coherence in groups will also increase their productivity. Consequently, it is important that we develop the tools and techniques needed to observe collective behavior and to identify patterns within it.

A number of factors can help groups work more coherently. These include standard procedures, prescribed design and development methodologies, required milestones and design documents, and the organizational culture, itself, with its projected expectations of behavior. These instruments can be effective for routine tasks, but they work less well for tasks that are unusual or that are particularly complex. For these situations, groups must rely on more flexible means. One term for a flexible, but coherent approach to a task is

---

[1]   I use the term *coherence* with regard to collaborative behavior to mean that the individual actions of a member or of a group contribute in a constructive and orderly way to achieving a desired goal. With regard to the artifact, coherence refers to the orderly and logical relationship of its parts.

strategy. Because this discussion is primarily concerned with groups doing nonroutine tasks for limited periods of time, rather than with ongoing organizations with established procedures, strategy is particularly important concept.

In the remainder of this chapter, I consider strategy as it applies to collaborative groups and, thus, to a concept of collective intelligence. I look, first, at the general concept of strategy. After that, I discuss strategy in individual work and then extend the discussion to groups. The goal of this last section is to identify a set of constructs and a methodology that can be used to recognize and understand strategic behavior in groups.

Concept of Strategy

Strategy is both a familiar and an elusive concept. We all have at least an intuitive sense of strategy, but it is hard to define the term precisely. One reason is that it is used to refer to at least three different things. *Abstract strategy* exists as a set of rules, heuristics, visual images, or other forms that can be applied to a large number of specific situations. Frequently this type of strategy is associated with a particular type of task, such as writing or programming; thus, one has a strategy for planning and writing a document or planning and writing a computer program. People who have developed an abstract strategy for a task usually have the sense that they will know how to perform that task when an occasion arises, and they are likely to have a ready set of procedures they can draw on to do so. Thus, although abstract strategy is often task specific, it is also general in the sense that it is not restricted to any particular instance of its use.

Although individuals or groups may develop abstract strategies on their own, frequently particular strategies are taught as part of teaching a skill. For example, many writing instructors teach a strategy called the *stages model*; it advocates that writers follow a particular sequence of actions — planning, followed by drafting, followed by editing. A similar approach to computer programming is called the *waterfall model*. Thus, abstract strategy is often prescriptive, indicating a particular procedure that one is urged to follow. Both the stages and waterfall strategies have been criticized as being too rigid (i.e., too linear), reflecting what an ideal writer or programmer might do, but not what most writers and programmers actually do (Hayes & Flower, 1980; Boehm, 1988). But this need not be the case. Abstract strategies can exist as flexible models that provide the person or group with varied sets of options from which to select. I will illustrate this form of strategic model in the next section by describing a strategy for writing that is based on a general graph structure, rather than a strict linear sequence.

A second concept of strategy arises when an abstract strategy is applied to a given set of circumstances, for example, when a person sets out to write a particular document or a particular program. These events frequently generate a concrete product — a *plan of action* — that records the result of the strategic process. Because a plan describes particular sequences of actions whose completion would accomplish the task, we sometimes refer to the plan, itself, as a strategy. An outline for a document is an example of a plan of action, because writing all the various sections identified in the outline will, ultimately, produce the document as a whole. PERT charts and CPM diagrams are more temporally oriented schema and can describe parallel sequences of actions that can go on independently.

A third concept of strategy refers to *patterns of behavior* exhibited by an individual or group. Imagine looking down on someone carrying out a task. The person's behavior may at first be incomprehensible, but, in time, we would begin to recognize patterns and steps in his or her behavior. When this happen, we might describe those patterns in terms of the strategy we infer the person to be following. For example, chess matches between expert players are often described this way when a player is thought to be following a well-known strategy, such as Ruy Lopez or the Sicilian Defense.

If we step back and look at all three concepts of strategy, we can see that they are closely related. Abstract strategy applied to a given task produces concrete intermediate products, that is, a plan of action. Abstract strategy also produces patterns in the observable behavior of a group and its individual members. Thus, understanding the abstract strategies collaborative groups use is key to understanding the behavior of groups and the ways they go about constructing artifacts.

The information flow model illustrated in Fig. 2.1 made distinctions among tangible, intangible and ephemeral forms of information. Tangible products were further divided into target and instrumental, whereas intangible knowledge was divided into private and shared knowledge. Differences in the behavior of groups can be characterized in terms of differences in the information products they produce. For example, some individuals or groups will routinely develop particular instrumental products before they develop target products; others will bypass or spend less time on instrumental products and move more directly to target products. Thus, different paths through the framework point to different strategies being followed. As we try to draw an individual's or a group's strategy into sharper focus, we should look closely at the various transformations suggested by the flow model and at the evolving artifact in which those changes are visible.

The distinction between abstract and observed strategy is similar to the distinction between language competence and language behavior. Individuals are thought to possess a vocabulary of words, a semantically structured memory keyed by words in the vocabulary, and a grammar. When confronted by particular situations and needs, they draw on these constructs to generate specific expressions that address those needs. We can imagine an analogous concept of strategic competence. An individual or group possesses a vocabulary of actions, a long-term memory that stores information about the task content domain, and a strategic grammar. When confronted with a particular task or problem, they generate specific sequences of behaviors that address those needs. Although we seem to be born with the basic apparatus that underlies linguistic competence, we must develop the specific knowledge that comprise the vocabulary, memory contents, and grammar that enable it. Similarly, we are probably born with basic problem-solving and planning apparatus, but we develop specific methods and behaviors and learn how to use them through experience and training.

In the rest of this chapter, I consider how we might go about uncovering the strategic grammar that underlies coherent collaborative behavior. In doing so, I first consider strategy with respect to individuals, then groups.

Implicit in this discussion is both a paradigm for research and a long-term agenda. As emphasized in chapter 4, I suggest that we begin with observable data and concentrate, first, on building analytic and descriptive models of strategy. Expressed as grammars, those models will enable us to parse behaviors and produce structural descriptions of them. Those descriptions, in turn, will enable us to make comparisons between individuals working alone and within groups, to test hypotheses about what does or does not affect strategy, and to see whether or not differences in strategy make a difference in productivity or quality of work. Thus, valid analytic models would produce a number of practical benefits. Over time, they should also account for finer and finer grains of behavior and incorporate an increasing number of situational factors. That detailed knowledge should eventually provide the basis from which to infer the mechanisms that underlie observable behaviors, leading to generative models. Generative models should, in turn, lead to detailed predictive studies. Without a prior base of knowledge built from studies that use analytic models, predictive models are likely to be able to address only small, disjointed behaviors, with no realistic prospect of ever getting to the really hard and interesting problems found in the work of actual collaborative groups.

Individuals

I begin with individual strategy for two reasons. First, individual work comprises a large fraction of most collaborative projects. Second, although models of individual strategy are not simple, they are more tractable than detailed models of collective strategy will be. If we can build the first, we may be able to extrapolate from them to handle the more complex strategies of groups.

The discussion should be understood within the terms and constructs introduced earlier in chapters 4, 5, and 6. I assume that a complex conceptual construction task is being carried out by an individual working in close conjunction with a computer system, preferably one designed to amplify the user's intelligence with respect to the given task (or set of similar tasks). Consequently, thinking will occur largely in terms of a succession of computer-mediated cognition cycles. The strategy followed by this user can be described in terms of a strategy that applies to an individual processor for tangible knowledge. Because that processor operates in close conjunction with the artifact that serves as a form of long-term memory, the results of the user's strategy will be visible in the succession of small-grain changes made to the artifact. Thus, the primary issue is uncovering patterns that occur within the sequence of operations that produce those changes and the rules and other factors that are responsible for them.

Because the discussion is oriented toward analytic or descriptive models, we must, first, consider the *data* — the sources and types that reflect strategic behavior and that will be analyzed by the model. This is important because the data represent all of the information available to the model with respect to a given instance of a task. Second, we must consider the analytic *framework* that will be used and the terms and categories it provides. Third, we must consider the *formalism* in which the model will be expressed.

To make the discussion concrete, I use expository writing as the illustrative task. Thus, the discussion is oriented toward the strategies writers use to plan and write technical or scientific documents. As the example system, I use the Writing Environment (WE) described in chapter 4. The discussion also draws on the mode-based model of writing discussed there. Focusing on writing strategy, rather than strategy in general, is a convenience; the approach and concepts apply to a broad range of conceptual construction tasks.

Data

Because the data strongly influence what the researcher can infer about the user's cognitive behavior, choosing the right kind of *data* — at the right level of granularity and with the right descriptive parameters — is crucial. Several options are available. In this section, I briefly discuss four kinds of structured data, or *protocols*, that have been used to study strategy as well as other cognitive and human-computer interaction issues. These include *concurrent think-aloud, events, video,* and *action protocols.*

Concurrent Think-Aloud Protocols. This method was developed by Newell, Simon, and others at Carnegie Mellon University during the 1960s to study complex, problem-solving behaviors (Newell & Simon, 1972). It produces a written record of subjects' trains of thought based on the subjects' own verbalizations of their thinking while they perform the task being investigated. Tasks that have been studied using think-aloud data include writing documents and computer programs, solving arithmetic problems, assembling physical devices, playing chess, and, more recently, using various computer systems. Under laboratory conditions, subjects are prompted to continuously narrate their thoughts; under naturalistic conditions, such as a subject writing a paper at home, prompting is impractical. Consequently, think-aloud protocols often differ significantly in the levels of detail reported by different subjects under different conditions.

Think-aloud protocols have been debated on several grounds (Nisbett and Wilson, 1977; Ericsson & Simon, 1980). Subjects do not always know what they are thinking. The act of thinking aloud can, under some circumstances, distort thinking. The data are hard to encode in a consistent manner and doing so requires considerable time and effort, making them expensive. And, as noted previously, subjects produce protocols that differ significantly from one another in completeness and detail. In spite of these problems, think-aloud protocols can, when used judiciously, provide richly detailed insights into a person's motives and intentions not available from other types of data.

Events Protocols. To address problems of cognitive interference and the laboriousness of manual preparation associated with verbal protocols, some researchers have used the computer system itself to collect data. For command-driven systems, data is collected in the form of keystrokes (Card, Moran, & Newell, 1983). For direct manipulation systems, protocols can be collected for a wider variety of input and display events, such as movements of the mouse and graphic information displayed on the screen. So-called X-Windows protocols are an example of this type of data. I refer to both keystroke and low-level direct manipulation protocols as events protocols.

Events protocols solve several problems raised by think-aloud methods, but they also raise others. Because they are recorded by the system, they do not require initial manual input. However, because the data are so fine-grained, it is difficult to infer what the keystrokes or mouse events add up to in terms of system commands, and inferring their effects on the conceptual structures being developed by the users may be impossible. Consequently, before conceptual or cognitive analyses can be done, events data must be manually categorized by a trained analyst, raising problems of consistency and costs.

Video Protocols. This method has been used alone as a supplement to both think-aloud and events protocols. Subjects are video taped as they perform a task. For studies involving a computer, these data can show what a person is doing when not thinking aloud or interacting with the computer. They can also show what is displayed on the screen — information not available through think-aloud or keystroke records. Thus, a major benefit of video data is that users' behaviors are captured in a more complete context (Mackay, Guindon, Mantei, Suchman, & Tatar, 1988). However, video protocols require extensive analysis and coding, raising similar problems of consistency, time, and labor.

Action Protocols. A fourth type of protocol data records the user's actions with respect to a system's data objects and control functions, rather than low-level keystrokes or mouse events (Smith, Smith & Kupstas, 1993). Action protocols record movements of the mouse from one window to another — rather than from one screen coordinate to another — and selections of identifiable data objects — versus simply reporting that the mouse was clicked at a particular screen location. They also report menu options selected and character strings typed in as names or labels for objects. Thus, action data are far richer than events data in terms of the semantics of the task.

Currently, the approach is limited to studies that involve computer systems for which the researcher has access to the source code or can persuade those who do to make the necessary modifications. Nevertheless, action protocols offer many advantages over other forms of data. Like events protocols, they solve the problem of cognitive interference raised by think-aloud methods by being passive and unobtrusive; and because they are recorded in machine-readable form, they eliminate the need for manual transcription, required of both think-aloud and video protocols. But unlike events protocols, action protocols provide data in which content objects are identified, enabling analyses at a conceptual or cognitive level without manual coding. Because they represent a more abstract level of activity with respect to the task, they are an order of magnitude less numerous than events protocols. Although not a panacea, they offer a number of advantages that make them especially well-suited for studies of strategy. Consequently, they figure prominently in the discussions that follow.

Action protocols are normally recorded one action per record. Each record includes several parts. A symbol identifies which of a designated set of actions has occurred; for example, *create_node* might indicate that the user has created a new node on the display. Each record also includes the time the action begins, its duration, and parameters relevant for the particular action, such as the identification number of the node and its location on the display, or the character string typed in. Thus, different actions have different parameters. For the Writing Environment, the set of designated actions numbers approximately 50. We can think of this set of identifiable actions as a vocabulary of action *types* from which individuals working with the system can select.

As a users works with a system that is recording action protocols, he or she generates a sequence of action records. Each action within the flow of work can be viewed as a *token* of its corresponding action type. In a language, words are normally organized into sentences or expressions that serve some purpose. Similarly, actions are normally used to accomplish some purpose with respect to the task. If we are to understand a user's strategy, we must understand what these sequences of actions are telling us about the user's intentions with respect to the intellectual construct he or she is creating. One way to approach this problem is to build a model of strategy in the form of a grammar that can be used to parse expressions in the language of action protocols, just as conventional grammars can be used to parse expressions in English, French, or Pascal. In the sections that follow, I describe how this can be done.

Framework

To build a model of strategy, we must first decide on an *analytic framework*. This is an important decision because it includes the terms and categories in which the model will be defined. Ultimately, this set of categories is arbitrary, because one can always construct alternative categories with which to describe a given set or type of data. However, because the framework in which the model is built provides the mental constructs in which we think and talk about strategy, it should be given some care. First, the categories within the framework should denote the "seams" in the phenomena to be considered, the "natural" breaks in the data. Second, we should think of these categories as falling within different levels of abstraction. If the model takes the form of a grammar, the categories will constitute the nonterminal symbols in the grammar, analogous to terms such as *noun* and *noun phrase* in natural language grammars. Thus, lower level symbols, denoting fine-grained behaviors, will be composed to form higher level symbols, denoting more extensive behaviors; conversely, higher level symbols will be decomposed into finer grained symbols. Finally, because category names are unlikely to be original, they will carry associations with other models; we should be sure that such associations are intended and reinforce desired relationships.

In chapter 4, several analytic frameworks were discussed. The GOMS framework included four components: goals, operations, methods, and selection rules. As emphasized there, the selection rule component requires that GOMS models be predictive models. They are capable of handling short, independent sequences of actions that occur over durations of a few seconds to 1 or 2 minutes; but they are not currently capable of handling sequences that take place over extended periods of time

Consequently, I discussed an alternative framework, based on cognitive modes, that is better suited for analyzing and characterizing strategic behaviors that extend over several hours. Conceptual behavior is viewed as a sequence of cognitive modes, each composed of different combinations of goals, products, processes, and constraints. I do not repeat that discussion here, but for the remainder of this section, I use the categories included in the modes framework as the building blocks with which to construct models of strategy.

Formalisms

The third major component needed to build strategic models is a *formalism* in which to describe relationships among the terms and categories of the analytic framework. Models have been developed in many different formalisms, including algebraic expressions, set theory, production rules, and graph theory. Although the choice of formalism is ultimately arbitrary, some are inherently more powerful than others, and some are better suited for dealing with particular types of data and the phenomena they represent. I focus on strategic models that take the form of formal grammars. However, a number of different options are available for expressing grammar rules. I discuss two: *production rules* and a graph-based form, called *augmented transition networks* (ATN). My colleagues and I have developed several different strategic models for writing using these formalisms, which I draw on for illustrations. But, I want to emphasize, the formalisms are general, and other models could be built within these same formalisms by defining alternative sets of rules.

WE Production Rule Grammar. Our first grammar was expressed as a set of production rules. Production rules denote a relationship between a single symbol on the left side of the expression and a pattern of one or more symbols on the right (Newell & Simon, 1972). The pattern may be a sequence of symbols or a Boolean expression, such as an *and* or an *or* expression. A rule is normally interpreted as a mapping between the two sides. If it is used in a recognition procedure, an occurrence of the pattern on the right within an expression can be replaced by the single symbol on the left. Conversely, if it is used in a generative procedure, the symbol on the left can be expanded and replaced by the more detailed pattern on the right. For example, the rule, $S ::= NP + VP$, says that the symbol, S, can be replaced by the pattern, $NP + VP$. More detailed rules might map NP onto expressions that involve nouns — the class of which might be denoted N — and, in turn, N might be mapped onto a set of specific nouns — such as *dog* or *cat*.

When used with a parsing program, the rules map specific expressions onto a parse tree that denotes the abstract grammatical structure of the expression. Conversely, when used with a generative program, the rules produce valid expressions in the language defined by the grammar. One way of looking at the parse tree is that it represents the sequence of rules that could be used, starting with the most abstract symbol in the grammar, to generate the expression that was parsed.

Production rule grammars are context free. Because studies involving computers must often take into account the context of a given behavior, the formalism must be extended if it is to provide this capability. The WE production rule grammar included some half-dozen functions that recognize contextual relationships among the data. For example, they recognize the particular type of graph object involved in an action, such as disconnected sets of nodes versus connected graphs; they also recognize distance relationships between data objects, such as the distance between two nodes on the screen. Consequently, the same user action may be interpreted differently under different conditions. For example, the WE grammar infers that a different cognitive operation has occurred when a user crates and places a new node close to an existing node on the screen, versus placing it in a location with no other nodes nearby. (Smith et al., 1989).

The nonterminal symbols in the WE production rule grammar form a hierarchy. The start symbol is a *session*. It is composed of a series of cognitive modes in which a sequence of cognitive processes are used to produce particular cognitive products or changes to products. The top four levels of the grammar describe the cognitive behavior of the user from a modal point of view and are general with respect to the task.

The bottom two levels link the cognitive portion of the model to a specific writing system — WE. To transfer changes in cognitive products, which take place in the head, to the representation of those products, within the computer, the user performs a series of system operations, each composed of one or a short sequence of system actions. Actions are the protocol data recorded by the system, as described previously, and constitute the terminal symbols of the grammar. Operations represent a more general view of the user interface. For example, the *create_node* operation requires four user actions: designating a screen location with the mouse, highlighting and then selecting the create node option from a pop-up menu, and typing a name or label for the node.

The WE production rule grammar is illustrated in Fig. 7.1. The figure is divided into two parts. At the bottom is a partial representation of the user's display. At the top of the screen is a portion of the grammar's parse of the user's actions that produced the display. The illustration is focused on the actions that resulted in the creation of the node labeled *n*. Because the protocol for the entire session is available, we can look down on it and see both forward and backward in time. In this case, the user has recently created nodes *n-1* and *n-2*, in two previous operations, and will soon create nodes *n+1* through *n+4*. Five of these seven nodes are interpreted by the grammar as forming a *cluster*, whereas two — nodes *n+3* and *n+4* — are interpreted as "solo" nodes. This distinction is based on the relative distances between the various nodes, determined by the grammar's context recognition functions.

In the portion of the parse tree shown at the top of the figure, we can see the actions that resulted in the creation of node *n* as well as the grammars interpretation of the user's cognitive behavior that resulted in nodes *n+1* through *n+4*. To create node *n* requires the four system actions, shown at the lowest level of the parse tree — opening a menu, highlighting and selecting the create node option, and typing a name for the new node (i.e., "Introduction"). These actions comprise the system operation, *create_node*, shown in the second level of the tree.

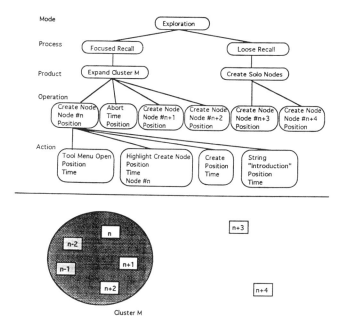

Fig. 7.1. Illustration of Writing Environment (WE) production rule grammar, showing a sequence of actions and their parsed interpretation by the grammar. Note that the same *create_node* operation can indicate different cognitive processes, depending on context or position of the node relative to other nodes.

Within this particular context, this operation is interpreted as an indication that the cognitive product, *Cluster M*, has been expanded (level 3), as a result of a *focused recall* cognitive process (level 4). We can also see that the user's activities that produce nodes $n+1$ and $n+2$ are interpreted similarly. However, the operations that produced nodes $n+3$ and $n+4$, by virtue of their locations on the screen, are interpreted differently — they indicate that a *loose recall* cognitive process has occurred. Thus, the same system operation may be interpreted differently by the grammar, supplemented by the recognition functions, depending on contextual factors such as the location on the screen where a particular operation takes place. Finally, all of the activities involved in creating nodes $n-2$ through $n+4$ are interpreted by the grammar as occurring within an *exploration* cognitive mode. The illustration stops at this level; the overall session is comprised of a number of such modes.

The WE production rule grammar has been used to analyze some 5 user protocols, each representing 2- to 3-hour sessions recorded in our laboratory under seminaturalistic conditions. Results of these studies have been described in Lansman and Smith (1993); Smith and Lansman (1992); Lansman (1991); and Smith and Lansman (1989). Figure 7.2 shows the strategic behavior of two subjects as they planned, wrote, and revised short, 3–4 page documents. On the left side of both displays is a taxonomy of actions defined by the researcher; actions have been organized into three categories — *explore, organize,* and *write.* On the right are frequency distributions for each category as well as each operation within a category. Time extends from left to right. Individual operations are represented by tick marks at the time each begins, and a horizontal "tail" on the tick mark indicates its duration.

The computer tool used by researchers to produce these displays allows them to divide a user's session into segments, indicated by vertical lines that extend from top to bottom in each display. In the figure, Subject 1's session has been divided into four large segments, and Subject 2's display has been divided into two segments. Histogram at the bottom of each display show the total number of operations for each category of operations for each such segment.

The data, grammar, and display tools let us literally see users' strategies as they are reflected in patterns in their cognitive behavior that extend over several hours. Subject 1 first created a group of exploratory products. He then constructed the top of his tree and wrote blocks of text for each node in the tree. Finally, he went back and filled out the bottom of the tree and wrote text for those nodes. Thus, the strategy followed by this first user is almost a classic stages or waterfall model in which the one first plans, then writes, and finally, edits the draft.

This strategy is markedly different from that of Subject 2, who characteristically created a node, in exploration mode, and then immediately wrote a block of text for that node, in writing mode. This writer constantly moved back and forth between the two modes, rather than spending substantial time in one before moving to the other. The same pattern predominated in the second part of the session, but between organizing and writing modes.

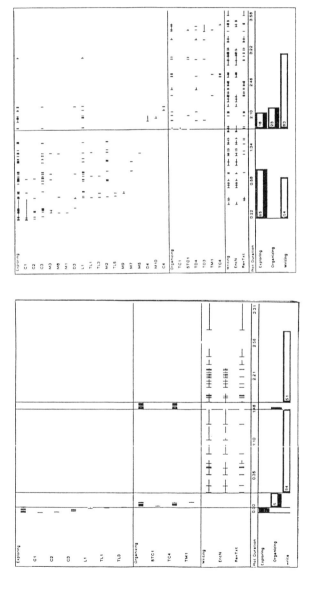

Fig. 7.2. Two writers' strategies. One shows a pattern of sustained planning followed by sustained writing; the other shows a continuous movement back and forth between planning and writing.

One possible explanation for these very different strategies is that the first subject may be more abstract in his or her thinking whereas the second is more verbal. The first strongly separated brainstorming and planning activities from linguistic encoding; for the second, brainstorming and planning were consistently interleaved with actual writing. Additional research is needed to understand these differences in strategy and their causes and effects. The point to be emphasized here is the access to users' strategies provided by this methodology and its associated tools.

WE ATN Grammar. The production rule grammar described previously works satisfactorily, but the architecture of the computer program in which it was implemented is awkward, making refinement of the model difficult. In addition, the recognition functions that provide context sensitivity are not well integrated into the grammar. To solve these problems and to extend our thinking about strategic behavior, we have defined a second grammar, using the Augmented Transition Network formalism (Woods, 1970). Although this formalism has been used primarily in natural language understanding systems, ATN grammars are well-suited for analyzing human-computer interaction data.

An ATN grammar is expressed as a set of graph structures whose nodes represent *states* and whose links represent *transitions*. The states are arbitrary and abstract, whereas the descriptive labels attached to the links/transitions denote the nonterminal symbols of the grammar. *Tests* may be attached to the transitions, providing a convenient mechanism for checking contextual parameters, and values may be written to *registers* by one part of the grammar and later accessed by another. ATN registers, which are simply global variables, are not to be confused with computer hardware registers.

A parse is performed by a program that reads a sequence of symbols from a language and, for each symbol, traverses the ATN graph structures, checking for conditions, performing tests, and recording information in registers. A traversal of a link in one graph may also result in a call to a different graph; when the traversal of the latter graph is complete, control returns to the link/transition from which the original call was made. Of course, not all transitions will be successful, and the parser must be able to backtrack when it reaches a dead-end and try an alternative path. The parse is complete when the complete sequence of symbols has been read and the parsing program has reached a special *stop* state. After the parse has been completed, a parse tree that represents the grammatical structure for the string can be constructed from the labels along the path of successful transitions that led to the stop state.

Our decision to use ATNs was motivated by several factors. Its register and test capabilities make it convenient to incorporate tests on parameter values — such as the spatial coordinates of a node — into the rules of the grammar. This allowed us to incorporate the same functionality provided by the context recognition functions used by the production rule grammar directly into the ATN grammar. Second, because ATNs have the power of a Turing machine, no other formalism is more powerful. Thus, ATNs offer an attractive formalism for studies of strategies.

To gain a feel for the ATN model, consider Fig. 7.3. The model of writing strategy defined in the ATN grammar differs from the production rule model in several ways. Most obvious is that the ATN model includes a varying number of levels. In the production rule grammar, all sequences of actions were interpreted in terms of a six-level tree, ranging from session and modes at the top to operations and actions at the bottom. In the ATN grammar, some modes are "deeper" than others. A second distinction is that the computer system, itself, has been incorporated as a fundamental part of the task model. This is done by including the user's attention to the tool — manifest as a sequence of actions that address the system rather than the conceptual artifact being developed — as a mode of thought analogous to substantive modes — such as exploring or organizing.

At the highest level, users are assumed to be *working*, as indicated in the first graph. For whatever reason, they decide to write a document or to continue work on a document they were working on earlier. To do so, they engage a strategic mode that contains a model of the writing process. In the grammar, this is represented by beginning a traversal of the link marked *Write Document*, which generates a call to a separate ATN graph having the same label. When all activity associated with this graph is completed (including any additional calls to lower level graphs), the traversal of the link in the top graph will be completed and the grammar will stop. This pattern of descending and ascending by starting to traverse a link, calling a lower level graph, completing its traversal, and returning to the higher level graph to complete the traversal of the original link is repeated throughout the ATN grammar. Let's now follow one possible path through the grammar down to a terminal symbol.

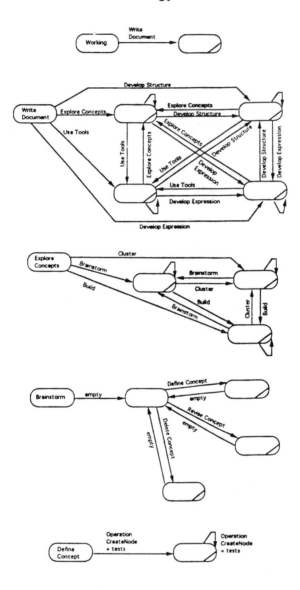

Fig. 7.3. A portion of the Writing Environment (WE) Augmented Transition Network (ATN) grammar.   It shows a possible vertical path through the grammar, from initial intention to write, through brainstorming mode, to the define_concept action.   The path could appear as a branch in the parse tree for a user's session in which a define_concept action occurred.

Within the context of the Write Document graph, users can explore concepts, develop the structure of the document, develop its expression, or they can consciously address the tool. Assume that our writer decides to explore concepts. This produces a call to the Explore Concepts graph. In the cognitive mode associated with this graph, one may brainstorm, build clusters of ideas, or build component structures. Each choice leads to a lower level graph for the corresponding activity. If the user decides to brainstorm, resulting in a call to that subgraph, he or she can represent a concept (define it), revise an existing concept, or discard (delete) an existing concept from further consideration. Defining a concept is viewed as a basic cognitive process in the model. As the final graph structure shows, the define concept operation and the change in the artifact it produces are subjected to tests. If the conditions are satisfied, the transition in this lowest level graph is completed. Presumably, the user then goes on to another cognitive or conceptual process that is reflected in the next action in the protocol sequence. If the test fails, the grammar backs up and tries an alternative path through the ATN network.

Basic cognitive processes, such as the define concept process, are examples of the computer-mediated cognition cycles discussed in chapter 6. Thus, the rules and tests in the grammar that recognize these elements map between basic cognitive operations, which occur in the head and produce changes in the structure of concepts active in human short-term memory, and basic system operations, which produce changes in the data structure represented in the display. The particular cmc cycle — *define_concept* — used for purposes of illustration here can be seen in Fig. 6.3. Thus, the grammar maps patterns of user behavior as they are manifest in observable system actions to low-level cognitive processes, which, in turn, are linked through several levels of abstraction to the session as a whole. As a result, high-level behaviors, which may last for as long as several hours, can be decomposed into sequences of lower level behaviors, which last for only a few seconds. Conversely, lower level behaviors can be composed to form higher level behaviors, at several levels of abstraction.

In this section, I have outlined a methodology for observing and analyzing the strategic behavior of individuals carrying out complex, conceptual construction tasks, such as writing a document. These steps included observing users' behaviors and obtaining suitable protocol data, selecting an analytic framework that includes the categories in which the model will be cast, and identifying a formalism in which to define relationships among those terms.   Action protocols were selected because they provide both fine-grained data and sufficient contextual information to support a limited form of context-sensitive analysis.   The cognitive mode framework was used, for reasons discussed in chapter 4. Two formalisms were discussed — production rules, supplemented by context recognition functions, and Augmented Transition Networks.

Collaborative Groups

In this section, I extend the discussion of strategy from individuals to collaborative groups. The fundamental problem faced by collaborative groups is, on the one hand, to divide their work into semiautonomous tasks so that they can take advantage of the parallel efforts and the individual expertise of its various member and, on the other hand, to synthesize their respective contributions to form a coherent whole. The problems we face in trying to study and/or understand collaborative behavior is to observe these various activities, to construct analytic frameworks that fit the data, and to relate the categories and terms of the framework to one another to form analytic models.

As in the preceding section, I assume that a complex knowledge-construction task is being carried out and that much of the work is done in conjunction with a computer system. In this case, the work is being done by a collaborative group, rather than an individual, and the group is assumed to be using a collaboration support system, such as ABC. We can gain important insights into collaborative strategy by observing groups use the system to build the artifact, similar to the approach outlined in the preceding section for an individual writing a document. However, because many group activities do not directly address the artifact, the methods we use to collect data, to analyze it, and to model strategic behavior must also address activities, such as meetings, in which intangible knowledge is developed.

The approach taken for individual strategy was to focus on the individual processor and the various computer-mediated cognition cycles that operate within it. To consider collective strategy, we must extend the discussion from individual processor to the multiple independent processors that function within a group. We must also include the processor for shared intangible knowledge and the hybrid processor. In a collaborative group, all three forms of processing will operate at one time or another and, for larger groups, often at the same time. Thus, a concept of collective strategy must recognize patterns not just within the individual instances of these processors but also in their interactions with one another.

The discussion begins with the data. The issue is collecting and managing fine-grained data comparable to that available for the individual processor but extended to the other types of processors. Second, I consider a basic framework that provides categories and terms in which to describe the social as well as conceptual behaviors of collaborative groups. Finally, I consider a formalism that can be used to build analytic models of collective strategy.

Data

The goal is to collect data over the duration of a project that will enable us to both examine the fine-grained behavior of the group and see its overall strategy. To meet these requirements, we must record the behavior of all three types of collective processors at the level of individual mediated cognition cycles. This situation is illustrated in Fig. 7.4. Three members of a group are working independently, shown on the left of the figure, while three others are working together in a computer conference, shown at the top. Meanwhile, we can see five additional members holding a meeting, at the bottom of the figure. Also shown are two observers — one observing and taking notes on the meeting, the other doing the same for the computer conference. Let's look more closely at each of these three types of activity.

Multiple Independent Processors. Group members working independently at their respective workstations, shown on the left of the figure, comprise a set of multiple independent processors. As they work with a collaboration support system — for example, the ABC system — the system records each action they perform. Thus, each users' ABC session produces a protocol similar to those described previously for the WE system. Each such protocol is identified by the group member who produced it, the time it was produced, and other

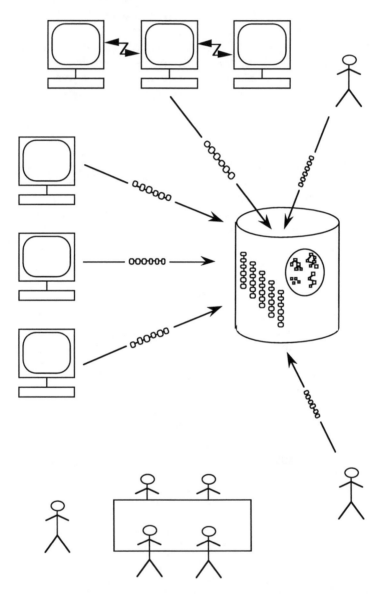

Fig. 7.4. Multiple streams of protocol data for group members working independently, in meetings, and in computer conferences.

similar information, followed by the sequence of records that identify the user's individual actions. Protocol data are stored in the same hypermedia data storage system that holds the artifact, shown at the right of the figure. Each such protocol, and the actions that comprise it, is suggested by an arrow; several prior protocols, along with the artifact, are also visible in the database.

Processor for Intangible Knowledge. Data that record actions that take place in meetings and other similar situations must be recorded and coded by a human being, preferably a trained observer who can identify different types of actions — that is, group-mediated cognition cycles — in a reliable and consistent manner (Cain & Reeves, 1993). Manual coding is needed whether the observer is actually in the room when the activity takes place, as suggested in this illustration, or views a videotape of the activity and codes it later. Individual behaviors are coded as to which of a given set of behavior types has been observed, the time the behavior takes place, and relevant parameters. This type of protocol has the same basic form as the machine-recorded protocols, but differs in the specific actions identified. Consequently, these data can also be kept in the same storage system and many of the same tools can be used with them. An example of a study based on this type of observation is described in the next section; a more thorough discussion can be found in Holland, Reeves, and Larme (1992).

Hybrid Processor. Data is also needed for the 1xn hybrid processor, shown at the top of the figure. In this example, three members of the group are participating in a computer conference in which they take turns working on the same part of the artifact. They can also talk to and see one another using supplemental voice and video channels. The protocol stream for the chairperson includes symbols that identify actions for all of the conference participants that affect the part of the artifact that is the subject of the computer conference, along with his or her actions before and after the conference. Actions of the other participants that take place in nonconferenced windows — representing individual actions that take place during the conference but are not part of the conference — are included in their respective protocol streams but not in the protocol stream for the conference.

The verbal and visual interactions that occur through supplemental communication channels are not included in the machine-recorded protocols. However, an analyst can observe the computer conference, including these channels, and take notes similar to those for meetings. These data can then be coded and entered into the database. Thus, as the figure suggests, a computer conference, when supplemented by video and/or voice channels, will (potentially) produce two protocols — one, machine-recorded; the other, manually coded observations. These two data streams can subsequently be merged or correlated with one another using the time parameter included in each record.

Thus, protocol data can be collected that include a record of all actions that affect the artifact and a substantial number of those that contribute to the development of shared intangible knowledge. Although not complete, these data are extensive and can provide a detailed understanding of a group's fine-grained behavior as well as broad, strategic patterns.

Framework

Earlier, I discussed cognitive modes as a framework in which to describe the cognitive behavior of individuals. Because the individual processor is part of both the multiple and the hybrid processors, cognitive modes are also relevant to collective behavior. However, to describe behaviors that take place in meetings and other situations where intangible knowledge is developed, we need a framework that incorporates social as well as cognitive factors. In this section, I will describe a construct, called a *mode of activity,* that does this.

Western psychology, with its emphasis on controlled experiments, has tended to focus on individual cognitive functions. Although this research has produced detailed knowledge of isolated processes, those processes have seldom been examined under real-world conditions. Thus, it has not addressed the quick oscillations between conceptual and social processes that occur in collaborative groups and lie at the heart of group-mediated cognition.

An alternative tradition, known as *activity theory,* developed in the first half of this century in the Soviet Union. Associated with Vygotsky, Leont'ev, and their followers, it assumed that mental processes are always situated in broader cultural and social contexts and should be studied within those contexts, rather than in isolation as required by controlled experiments.[2] Activity theory includes several key concepts that are useful for understanding collaboration and for building a concept of collective intelligence. These include *situated activity, mediating devices, higher* and *lower mental functions,* and the *zone of proximal development.*

Activity theory views all mental processes as being *situated,* because it asserts that individual cognition always takes place in, and is responsive to, socially created activities. Even when alone, individual thinkers typically interpret the issue at hand in relation to a mental activity learned from others. Thus, conceptualization is embedded in the culture, with respect to both the symbols and concepts that are the substance of thought as well as basic mental processes.

To explain how mental processes can be influenced by social factors, Vygotsky differentiated between what he called lower and higher mental functions. The key to this distinction lies in the role symbols play in abstract thought. Symbols function in the mental world as tools function in the physical world. Thus, symbols mediate between one's mental states and processes and one's environment. For example, the basic act of remembering, made possible by one's neural apparatus, is a "lower level" function; however, when people learn to use mediating devices as tools for remembering — such as mnemonic associations derived from their language and culture — their memory capacity is increased and they have more conscious control over memory-related processes. Thus, mnemonic-assisted remembering is a form of "higher" mental function.

Vygotsky argued that mediating devices are largely invisible, or *fossilized,* under normal circumstances, because once learned or developed, they become habitual and individuals are unaware that they are using them. Consequently, mediated cognition can best be observed when new abstract devices are being developed or when new technologies are being introduced, and before the new forms of cognitive behavior that will inevitably develop become routine. Thus, there is a narrow window of opportunity when new mediating devices can most easily be observed, often occurring as an individual or group attempts to resolve a problem or snag.

---

[2] This discussion is based on an earlier summary included in Smith et al. (1990) which, in turn, was derived from Holland and Valsiner (1987) and from Vygotsky (1962, 1978, 1987).

Finally, Vygotsky argued that before we can carry out a task by ourselves, we must first learn the skill in proximity to another person. New skills are usually learned during work episodes that involve at least one (relative) neophyte and at least one (relative) expert. As the neophyte's ability develops, the expert curtails his or her participation, leading to the development of higher mental functions in the neophyte. Vygotsky called this situation the *zone of proximal development*, or *zoped*, for short.

Let's look back at collaborative behavior through the prism of these concepts. First, activity theory asserts that all mental behavior is situated within cultural and social contexts and is affected by those contexts. For collaborative groups, this context normally includes the organization in which the group functions, the group itself, and the physical settings in its various activities take place. For example, an activity carried out in a conference room (e.g., a meeting) is different from one carried out in a hallway (e.g., a chance conversation), although both may contribute to the development of shared intangible knowledge. Consequently, because context affects behavior, the analytic framework in which collaborative behavior is to be considered should incorporate parameters that characterize the particular situation in which the activity takes place.

Second, mediating devices are crucial in both activity theory and in collaborative groups. In activity theory, the principal form of mediating device is the symbol; for collaborative groups, mediating devices are more diverse. The computer system used to develop the artifact is probably the strongest mediating device (i.e., computer-mediated cognition). However, the ideas voiced by other participants during a meeting that influence one's own thinking are another form of mediation (i.e., group-mediated cognition). Other common mediating devices include audio/visual equipment, the whiteboard, and metaphors used to explain abstract concepts. Thus, although symbols and language continue to function as mediating devices, the tools and systems used as well as the group, itself, also mediate collaborative thinking. Consequently, the analytic framework should include mediating devices as a basic category.

Third, the Vygotsky distinction between lower and higher mental functions can be applied to a group's collaborative skills. As groups learn to work together more effectively, the successive stages they go through may be considered forms of higher mental function. Indeed, the coherent, integrated behavior I have referred to as collective intelligence can be viewed as a form of *collective higher mental function*. Thus, it can serve as a goal groups should aim to achieve. One reason they have trouble achieving this level of performance may be because there are few, if any, "expert" groups a "novice" group can work with to develop its collaborative skills. The notion that one group can serve as a mentor for another less skilled group, creating a zone of proximal development for collaboration skills, is a training strategy worth considering.

We can now define a *mode of activity* framework that is based on the cognitive mode framework but includes several new components from activity theory. We should keep in mind that cognitive modes were defined with respect to individual cognition and, in the terms of this discussion, an individual processor. The mode of activity construct is intended to address collaborative behavior. Consequently, it must account for the multiple processor, the hybrid processor, and the processor for intangible knowledge that, together, comprise the collective processor. Because an individual processor is one part of the multiple processor, which, in turn, is a component of the collective processor, we should think of mode of activity as an extension of cognitive mode and, conversely, a cognitive mode as a subset or projection of a mode of activity.

A *mode of activity* is defined as a multilevel structure that includes configurations of six basic components: goals, products, processes, constraints, situation, and mediating devices. Let's consider this definition, first, as an extension of cognitive mode, and then, as a multilevel structure.

Like cognitive modes, modes of activity include configurations of goals, products, processes, and constraints. *Goals* represent the abstract intentions of the group. Normally, they result in the production of some tangible *product* or in the further development of shared intangible knowledge. Thus, the types of products developed and/or used by collaborative groups are more extensive than those for an individual carrying out a similar task. This expanded set of information types requires a larger set of mental *processes* to build and access them. These include computer-, conference-, and group-mediated cognition cycles. Finally, *constraints* shape the situational matrix in which specific modes function. They limit the processes that can be used in a given mode, but they also define what is possible. More important, they define the norms of behavior that give a particular mode its identifying characteristics. For example, a brainstorming mode will carry a different set of expectations than a presentation mode. Part of this difference lies in the structures of the two modes and in the different sets of processes included in them, but a more fundamental difference lies in their respective atmospheres — one is likely to be formal and restrained, whereas the other is likely to be more relaxed, freewheeling, creative.

Thus, modes of activity are similar to cognitive modes with respect to these four constituent categories, although each category includes a more extensive set of possible values. However, modes of activity include two additional categories — the *situation* in which the mode occurs and *mediating devices* that may be used within the mode. I can best explain these extensions through an example.

Many meetings include or are oriented around a presentation. Although they may comprise an entire meeting, presentations are more commonly one activity within a larger agenda. For example, a presentation is often preceded by other business and followed or interrupted by discussion. Thus, the presentation is one mode of activity within a sequence of modes. Presentations normally take place in some form of conference or meeting room where participants gather at the same time and same place. The situation will be different and, as a result, the behavior of the group will be different if the presentation is made by video or takes place in an auditorium. Presentations usually include some form of audio/visual equipment, such as an overhead projector, a slide projector, or a VCR. The behavior of the group — including their understanding of the material and/or their agreement with the ideas presented — is influenced by these mediating devices. Thus, for a presentation mode of activity, we can identify the following additional categories and parameters:

- *situation*
    - conference room
    - same time, same place

- *mediating devices*
    - overhead projector
    - whiteboard
    - slide projector
    - flip chart

We can see the subset/extension relationship between cognitive mode and mode activity by recognizing that in the earlier discussion of writing modes, a particular computer system — WE — was assumed; thus, it functioned as a mediating device but was not explicitly incorporated in the task model because it was a constant — all writers covered by the model used WE. Similarly, the situation for the studies cited was constant — our computer lab and the conditions under which subjects performed the task.

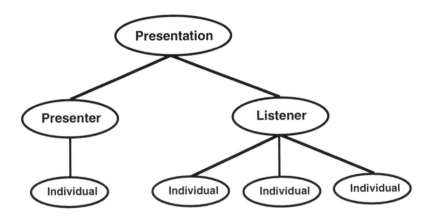

Fig. 7.5. Presentation mode of activity, comprised of one individual functioning as presenter and several as listeners.

The third extension is more fundamental. Modes of activity are multilevel structures. The mode in which a group as a whole is engaged is often comprised of several submodes in which subgroups and/or individuals participate. The multilevel structure for a presentation is shown in Fig. 7.5. At the top is the overall *presentation* mode, itself. It, in turn, is comprised of two submodes that operate concurrently — a *presenter* submode and a *listener* submode. At the bottom is a third level — the individual modes engaged by the participants. Each individual will presumably be involved either as a presenter or as a listener. An individual not participating in the mode of activity — for example, by daydreaming — can be viewed as having engaged a cognitive mode that is not part of the current group mode.

Other modes of activity will have different structures. But they all have at least two levels — the top one denoting the mode of activity for the group as a whole and the bottom one denoting the individual modes for the members participating in the activity. A variable number of intermediate submodes may be included, depending on the particular activity.

To show how the modes of activity framework can be used to characterize the behavior of a group, Kim Blakeley carried out an exploratory study during the summer of 1990. Over a 2-month period, she observed five meetings of a group during the early stages of a new project, taking detailed ethnographic notes similar to those described in the preceding section (Blakeley, 1990). Her analysis of these data identified three basic types of modes of activity: *discussion, presentation,* and *delegation.* Two of these modes, however, included variations on the basic form. In all, she observed the following seven modes or variations:

- Discussion
    - Discussion, in its basic form
    - Conflict-Resolution
    - Brainstorming

- Presentation
    - Presentation, in its basic form
    - Summary
    - Demo

- Delegation

Figure 7.6 shows the specific constituents that comprise these seven modes.[3]

---

[3] Blakeley's original table included only the four cognitive mode categories; situation and mediating devices were identified in the narrative description of each mode. I have updated the table to include these two additional categories of information.

Mode	Variation	Goal	Product	Process	Constraint	Situation	Mediating Device
Discussion	Discussion	• externalize information	• group-level awareness of information	• dialog • analysis	• take turns talking	• conference room • same time	• whiteboard
	Conflict-Resolution	• externalize information	• group-level awareness of information • resolving a conflict	• dialog • analysis	• take turns talking • one topic addressed • subset of group involved • different opinions encouraged	• conference room • same time	• whiteboard
	Brainstorm	• externalize information	• group-level awareness of information • generating ideas	• dialog • analysis	• take turns talking	• conference room • same time	• whiteboard
Presentation	Presentation	• introduce information	• group-level understanding of information	• teach and inform • listen, learn, question, evaluate	• one individual controls	• conference room • same time	• whiteboard
	Summary	• introduce information • receive same message	• group-level understanding of information	• teach and inform • listen, learn, question, evaluate	• one individual controls • carried out by leader	• conference room • same time	• whiteboard
	Demo	• introduce information	• group-level understanding of information	• teach and inform • listen, learn, question, evaluate	• one individual controls	• lab • same time	• computer system
Delegation		• assign task	• understanding of work responsibility	• delegate and explain • listen and evaluate	• senior member delegates	• conference room • same time	• whiteboard

Fig. 7.6. Seven Modes of Activity, including the processes, products, goals, constraints, situation, and mediating devices for each. (Adapted by permission from Blakeley, 1990.)

The description that follows is taken from Blakeley's summary and analysis of the group's second meeting,[4] following an earlier "kick-off" meeting for a new project. The purpose of the meeting was to identify a set of tools the group could use to help them build a new computer system. In these segments, we can see Blakeley's use of the modes of activity framework to characterize both the overall behavior of the group as well as the behavior of individual participants.

---

4   Blakeley italicized her direct descriptions of events; her discussion and analyses of events are shown in normal type.

The second meeting of this group took place on 5-23-90. The only co-leader present was George[5] (Sam was out of town). Bill and Paul were the only other faculty members present. Graduate students Fred and Tim also attended. The purpose of this meeting was to begin determining the software tools needed to build the collaboration system.

George opened by asking, "Who's here?" George immediately assumed the leadership role, but did ask if "anyone (had) an agenda for the day." When no one came up with an agenda, he announced that he would like to discuss the tools and applications needed to build the collaboration system.

George's opening comment reflected his concern in the first meeting about who would be joining the team. George's asking the group for an agenda served to give the other members a feeling that they shared control of the project. Perhaps this would further attract the individuals attending the meeting.

Fred was the first to offer a software tool needed in the new system. George went to the white board and began recording the information offered by the other group members. After several tools were listed on the board, George began checking off and circling items as an indication of those he thought important.

Thus, the group began the second meeting in a brainstorming activity. This activity was initiated by the leader as he solicited ideas from the group. Because the project was in its infancy and little was firm, the group seemed to be in an exploratory mode. Thus, the constraints on the discussion were few, resulting in a free-form flow of conversation among group members. Although George was soliciting ideas from the others, he was still in control of how these ideas were presented on the board. George was thus in a slightly different cognitive mode than the other group members. He was interpreting while the individuals presenting candidate tools were brainstorming.

The group as a whole was in a type of discussion mode, with the following characteristics:

---

[5] Not the real name of the participant;   all names were changed to protect confidentiality.

Mode type: Brainstorming

  Goal
  - to identify candidate tools needed in the
    collaboration system
  Product
  - a list of tools on the whiteboard
  Processes
  - Overall: active, balanced exchange of ideas;
    although not everyone participated, no one
    person did a majority of the talking
  - George: interpreting, writing
  - Others: brainstorming, offering items
  Constraints
  - general topic identified
  - low censorship with respect to relevance

George then reiterated that the group was "deciding what tools are needed to complete a project." Tim asked him to define "project." The leader answered, "Good point.   What kinds of projects should we aim for?"

Thus, the group was still not sure about the purpose of the collaboration system and was engaged in an ongoing process to define the system and its goals.   In addition to the goals of each mode of activity that the group went through, the group seemed to share a general goal of developing a common understanding of the project.

About 15 minutes into the meeting, Paul stated that it is important "not to confuse the possible multiple roles of the computer." He explained that "building things for the computer and using the computer to build are two different things, and we shouldn't confuse the two." George gave a short argument to Paul's statement. Paul's response to George was "possibly." George then drew a diagram on the board, showing where he thought the project should be on a scale between "design" and "CASE tools."

Thus, through the diagram on the board, Paul and George had an understanding of the other's concept, although they may not have agreed on the subject. This example shows how differences in opinion can be acknowledged and understood.   In a conflict-resolution mode, the individuals with the differing opinions debate until a common ground is located, or until positions are understood or left unresolved by agreement.   This particular scenario had the following characteristics:

Mode type: **Conflict-Resolution**

Goal
  • to further specify the purpose of the project
Product
  • a group-level awareness of the issue
Processes
  • dialog, debate
Constraints
  • a limited subset of the group participated in
    dialog
  • a particular topic was addressed
  • different points of view were encouraged

*As the meeting continued, Fred again brought up
the issue that he had introduced in the first
meeting: the goal of virtual proximity versus the
goal of collaboration augmentation. He stated that
a system with the goal of collaboration augmentation
would require more prerequisite cognitive work,
something that he indicated the "other group members
(were) more interested in." He also described that a
software system used to achieve virtual proximity
would be more easily accepted by users.*

Fred's comments again show that the group was
divided in its interests. Fred was on the "virtual
proximity" side of the house and was pushing the
project in that direction. Thus, coalitions of
individuals within the group were emerging. There
seemed to be two groups, each defined by a separate
goal for the collaboration project. Fred and George
were in one group, while Sam was separated from them
by his different goal for the project.

*The meeting continued with more discussion of
tools needed in the collaboration system. Fred
explained that with the new system, users should be
able to share documents although they may use
different text editors. He pointed out a benefit of
this as he said, "Today people get together and bite
the bullet and agree to use Tex even though they
hate it."*

*About 45 minutes into the meeting, George
presented a general summary of the meeting. He
stated, "This discussion has been useful to me and
has helped me understand the scope of what we are
really talking about here." He also summarized a
goal he had for the project.*

George's summary seemed to be an attempt to bring the meeting to a close. It also seemed to be an attempt to provide the other members of the group with the feeling that their time spent attending the meeting had been put to good use. Thus, the meeting had turned from a discussion to a summary with the following characteristics:

Mode type: Summary

Goals
- to bring the meeting to a close
- to assure group members of time spent profitably

Product
- a general description of the meeting

Processes
- George: summarizing, recalling, evaluating, talking
- Others: recalling, listening

Constraint
- presentation/statement made by the group leader

(Blakeley, 1990, pp. 28-31; reproduced by permission)

The meeting did not end at this point, but, rather, went through several additional mode shifts. These included another period of discussion followed by a second summary that did bring the meeting to a close. Fig. 7.7 shows the complete sequence of modes engaged during this meeting.

Blakeley's study demonstrates that ethnographic techniques can be used to observe in close detail a group's social and intellectual actions, that the resulting data can be analyzed in terms of an appropriate set of modes of activity, and that the overall strategic behavior of the group can be seen in the sequence of modes engaged by the group. As we attempt more comprehensive studies, the set of modes will have to be expanded to include behaviors other than discussions, presentations, and delegation. But the basic mode of activity framework, with its six categories of components, appears to provide a viable set of building blocks with which to develop models of collective strategy and to characterize the behavior of collaborative groups.

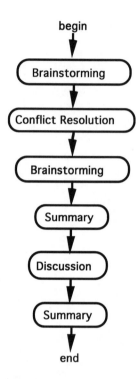

Fig. 7.7. Sequence of modes of activity observed during one meeting of a group. (Adapted by permission from Blakeley, 1990.)

Formalism

To understand the behavior of groups in a detailed way, we need to understand the interacting effects of its members with respect to one another and to the products and information structures they work with. Earlier, we saw that models of individual strategy could be built using the cognitive modes framework and the formalisms of production rule and ATN grammars. To model collective strategy, we need comparable, but extended, mechanisms that can deal with the more complex behavior of groups. In the preceding section, I suggested that we could use the modes of activity framework as the basic construct in which to define models of collective strategy. Here, I suggest a similar extension with respect to grammatical formalisms.

The system that implements a model of collective strategy will have to parse or otherwise analyze the multiple protocol streams that result from the different activities of the group. These protocols could be derived from individuals working alone, several members participating in a computer-supported conference, and groups involved in meetings or other similar activities. By analogy, imagine a natural language grammar that can parse not just a single input stream representing a single source of sentences but, instead, the multiple input streams representing all of the conversations going on in a room at one time, as might occur during a reception, including making sense of references in one conversation to remarks overheard in another. This is the nature of the task that an analytic model of collaboration must deal with.

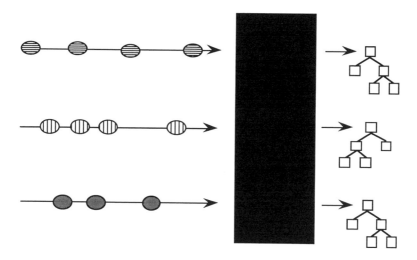

Fig. 7.8. N-dimensional grammar producing n separate parses. The grammar cannot identify effects one member's actions have on another.

To address these issues, our project is extending the ATN formalism to include rules that apply to multiple protocol streams. We refer to this formalism as an *n-dimensional grammar* to indicate its acceptance of *n* concurrent input streams, rather than the single steam of symbols accepted by a conventional grammar. N-dimensional grammars could take several forms. Fig. 7.8 illustrates one possibility. Three simultaneous protocol streams are input to the parser, represented by the shaded box in the figure. The parser applies its grammar to each protocol stream independently, producing three separate parses, shown as the three separate parse trees emerging on the right.

Rules for the three different types of processors — computer-, group-, and conference-mediated cognition — are written either as separate grammars or as an integrated set of rules, portions of which can be selectively applied to the three corresponding types of protocols. Thus, we could think of the grammar as three specialized grammars — one for each type of protocol — that have been yoked together so that they operate in tandem and as a single system. Following this approach, no attempt would be made by the grammar to infer the effects actions in one protocol stream have on actions in another. Instead, such inferences would be left to the researcher to derive from the analyzed data.

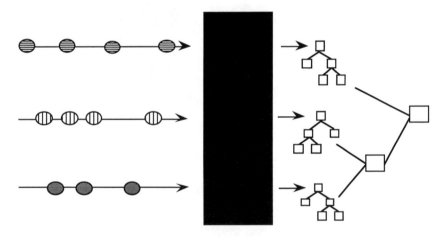

Fig. 7.9. N-dimensional grammar that produces n separate parses plus a group parse that identifies effects of interaction within the group.

A second, and more powerful, approach would include additional rules that identify patterns across multiple protocol streams. A parser based on this concept of n-dimensional grammar might still produce individual analyses for the different protocol streams — perhaps during a first pass through the data — but it would also produce a more general parse that spans all of the protocol streams, identifying interactions across individuals and subsets of the group. This second form of the grammar is illustrated in Fig. 7.9, in which three low-level parses have been further analyzed in relation to one another to produce a higher-level parse that represents the overall strategy of the group.

Although the first form of the grammar is appealing because of its relative simplicity, it is unlikely to be satisfactory, even for individual protocols. To see why, let's look more closely at some of the issues that must be addressed for each of the protocol types.

Multiple Individual Processors. Because the protocols for individuals working alone will be expressed in terms of computer-mediated cognition cycles and because the higher levels of the model can be expressed in terms of modes, we might expect this part of the grammar to differ from grammars that deal with individual behavior largely in the details of the task model — for example, a model for planning and writing computer code versus one for planning and writing a short document.[6] Similarly, the fact that the total system would have to accept multiple instances of this type of protocol would place additional requirements on the design of the parsing program but not on the grammar rules, themselves. Thus, at first glance, this part of the grammar seems similar to the grammars described previously that address individual behavior, that is, writers using the WE system.

However, this form of grammar cannot deal with the effects one member's actions have on another, either at the moment or at some future time. Momentary interactions for individuals working alone have to do with concurrency effects. If one member of the group is working on a particular part of the artifact and, at that very moment, another member wants to work on the same portion, under some conditions he or she may not be allowed to do so by the collaboration support system's concurrency control mechanisms. For example, ABC permits multiple members to read the same portion of the artifact, but only one member at a time may edit or change that part. If another user attempts to gain so-called write access to a segment for which another member already holds write access, he or she will be blocked. If a member asks for read access while another has write

---

6   To be consistent with other parts of the collective model, we would replace cognitive modes with modes of activity. This will require additional rules to take into account situation and mediating device parameters, but this extension would not change the fundamental form of the grammar.

access to that segment, he or she will be provided with a copy of the segment as it existed at the time the person with write access began work on the segment. Consequently, the reader's version of the segment may already be out of date if the writer has made a change but has not yet saved his or her version of the segment. Thus, the actions of an individual working alone on the artifact may be affected by the simultaneous, but logically independent, actions of other members of the group. A grammar whose rules apply only to a single individual could not accurately interpret situations such as this that result from the actions of another user.

Long-term interactions can take several forms. Of particular concern is the evolution of the artifact. If a group grants more than one member permission to write or change a given segment, although not at the same time, one member's work may be undone or changed by another's. For example, if member A adds a paragraph to a project document, member B could read that paragraph several days later, disagree with it, and delete or edit it. When member A comes back to the task, he or she would find his or her earlier work changed or missing. Because a parser that analyzes actions from an individual processor takes into account the effects of those actions on the artifact, it would find a discontinuity with respect to the data from one session to the next. Thus, it would have difficulty interpreting A's subsequent actions with respect to the artifact.

Issues such as these pose two different kinds of problems. First, they pose technical problems for the parser. If the grammar were built as three logically independent grammars, as illustrated in Fig. 7.8, it would include rules that treat such events as "black" events — they occur; the grammar produces a superficial analysis of the user's responses to them; but it could not develop a rationale for the user's subsequent actions based on the semantics of the black event. Let me illustrate this by continuing the first example. If a member of the group was denied access to a segment — either repeatedly or continuously for some period of time — that person would likely try to find out who else was working on that segment and then either arrange, informally, times when he or she could get access or initiate a computer conference so that they could work together on the segment. A grammar of the first type could not make an explicit connection between the access event and these responses to it. On the other hand, the second form of the n-dimensional grammar could include rules that do make connections between one user's working on a given segment and another user's attempts to gain access to that segment.

If the members of a project frequently encounter events such as these, it may indicate more fundamental problems. At the least, the situation suggests problems of inefficiency — individual work is being delayed or even destroyed. But it may indicate that the decomposition of the design and/or work assignment is faulty. It is precisely behaviors such as this that we would like a model of collective strategy to identify. A comprehensive perspective that looks down on the project as a whole and can see interactions such as these can only be gained from n-dimensional grammars of the second type.

Processor for Intangible Knowledge. The actions in this type of protocol occur in situations, such as meetings, where intangible knowledge is developed and consist of group-mediated cognition cycles. Consequential, they must be coded by an observer. The observer records individual behaviors and, if a mode of activity framework is being used for the study, the submode or role, such as presenter and listener, for the action. These behaviors can be recorded either as separate protocol streams, one for each participant and/or submode, or the respective actions can be interleaved to form a single protocol for the entire situation, with parameters identifying both individual and submode for each action. The two options are, of course, equivalent, because one can be mapped to the other.

If we think of the protocol streams as divided by participant, then the data look like that for multiple independent processors. An analysis will probably look for patterns of coherence within the behavior of individuals, in temporary coalitions, and in the group as a whole. Thus, rules will have to be included that focus on actions generated by each of these units. However, the grammar will also have to include rules that look at interactions across the various protocol streams, including behaviors influenced by submode or role. These are the same requirements as those just discussed for multiple independent processors, but the interactions are much stronger and more frequent, and the influences are more direct. Thus, although the rules for these two types of processors will be different because their information products and basic processes are different, the overall structure for both of these portions of the grammar will be similar.

1xn Hybrid Processor. Computer-based conferences can produce two protocol streams. A machine-recorded protocol recrods the actions produced by the various participants as they take turns working on the artifact. Rules that analyze this work will be similar to those for the multiple independent processors but will have to take into account the different origins of individual actions, because they can originate with different conference participants. Computer-based conferences may also produce a second protocol — a researcher's hand-coded observations of the social/conceptual interactions that accompany the conferencing group's work on the artifact.   This second protocol will resemble that of the processor for intangible knowledge, described previously, and rules for analyzing these data will be similar in form.   However, they will have to take into account the strong mediating effects of the communications channel(s) and the conferencing system, itself.   Additional rules will also be needed to identify patterns of interaction between these two forms of behavior and their respective protocol streams.

Although the rules for an n-dimensional grammar will be complex and their development will pose a significant challenge to researchers, a more basic problem lurks in the very concept of strategy when it is applied to large groups. A fundamental assumption with regard to individual strategy is that a person engages a single cognitive mode at a time, although he or she may rapidly shift from one mode to another. Thus, the person is brainstorming, planning, coding, and so on, but he or she is not brainstorming and coding at the same time. Consequently, we can describe a person's strategy by tracing his or her shifts from one mode or activity to another. This is not necessarily true for groups.

If the group is small — consisting of five or six members — the behavior of the group may be largely coherent. For example, a small group in the early stages of developing a new computer system might assign several members the task of going off and analyzing specific issues — such as the database system to be used or user interface requirements. After they report back, the group as a whole might then work out the high-level architecture for the system. If we analyzed this portion of the group's work, we might describe the group as experiencing a shift from brainstorming to early planning. Thus, although individuals or teams may engage in different activities, their individual actions come together at frequent intervals to support a single purpose for the group as a whole. Later, the group may shift to largely writing and/or coding activities. Eventually, we might expect to find the group primarily involved with testing and, perhaps, developing user documentation. Thus, although the work of small groups will not generally be as coherent as that of individuals, nevertheless, we may be able to identify the group's predominant mode of activity most of the time, while recognizing a diversity of behavior for individual members and/or subgroups. This was the case for the group writing congressional testimony, described in the second scenario in chapter 2. A study of design teams that found similar patterns of behavior is (Olson, Olson, Carter, & Storrosten, 1992)

The situation is more complex when the group is large. A project that involves 25–30 people is likely to include five or six such groups, as was the case for the software development project described in the third scenario in chapter 2. Although each individual team may exhibit coherent strategic behavior, it is not clear that the group as a whole will do so. Imagine a situation in which three teams are building three separate parts of a system, a fourth is building a set of analytic tools, and a fifth is studying the behavior of potential users of the system. Although the system- and tool-building teams must interact with one another to some degree, they may work independently for considerable periods of time. Similarly, the team concerned with human users may have only infrequent interaction with the others. Thus, at any one time, different teams may be at different stages relative to their respective parts of the project as a whole and, as a result, engaged in different modes of activity. What, then, is the overall strategic state of the group as a whole at any given time?

If it makes little sense to talk about a large group being engaged in a single overall mode of activity at any one time, we cannot expect to produce parse trees that describe the group's overall strategic behavior. In the parse trees described previously, a session was composed of the sequence of cognitive modes engaged by the subject while performing the observed task. If we took the same approach with groups, the penultimate level of the tree — or, if the grammar spanned multiple sessions, some high, but not penultimate, level — would consist of symbols that identify the overall mode of activity for the entire project at a given time. Because this will not be the case for large groups, we need a different way of representing collective strategy.

One possibility is to think of the strategy followed by a large group as resembling a multistrand cable, as suggested in Fig. 7.10. Each strand represents a single coherent activity extending over time, which runs parallel to the strands. Some strands represent individual behaviors, others collective behaviors such as meetings and computer conferences. When the activities of one group or one individual strongly interact with those of another, their respective strands merge for the duration of the interaction to form a single coherent, but

larger, behavior/strand.   When their activities diverge, the merged
strand separates back into parallel and independent strands.

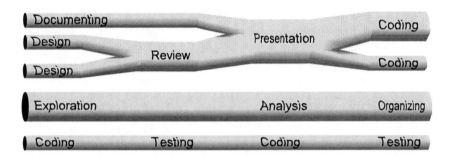

Fig. 7.10: The strategic behavior of large groups can
be thought of as multiple strands of coherent behavior,
parallel to one another in time.   Strands for individual
members merge during periods of collective work and
separate during periods of independent work.

In chapter 6, I suggested that IPS models of collective behavior
could be built more easily and more appropriately using a large-grain
distributed systems architecture rather than the Von Neumann
architecture used in models of individual cognition.   Within this
architecture, each strand of independent behavior within a group
would correspond to a separate process in the distributed systems
model of strategy.   Each represents an instance of one of the three
types of processors included in the collective processor — the
processor for tangible knowledge, intangible knowledge, and the
hybrid processor.   Thus, for example, when an individual member of
the group works alone on the artifact, a corresponding individual
processor in the distributed system model comes into existence for the
duration of work.   Similarly, analogous processes are launched for
meetings (and other similar situations in which intangible knowledge is
developed) and for computer conferences.

Each processor runs the parsing program and parses the protocol
generated by the corresponding activity in the group.   Because
processes can exchange messages in this architecture, the overall
system can interpret interactions between separate activities.   Thus, the
overall behavior of the group is modeled by the dynamic topology of
the distributed system and the flow of messages within it.   The
behavior of the distributed systems model, in turn, can be represented

by the multistrand cable, where each strand corresponds to a separate thread of control in the system and, hence, a separate coherent behavior in the group. Within each strand, detailed behavior can be represented by a conventional parse tree, because each such behavior is coherent.

Considerable research will be needed to build distributed systems models of strategy as suggested here, the n-dimensional grammars that run in its processes, and display programs that allow researchers to visualize the results of the model. I suspect, however, that any approach that is simpler will miss much of the complexity and interaction that occur within large collaborative groups and, hence, be of limited value.

To summarize briefly, I have discussed three components needed to build models of collective strategy. First, data must be collected that can provided a comprehensive view of the fine-grained activities of a group. These data must capture the actions of individual members working at their respective workstations, their interactions with the system and with one another during computer-based conferences, and their behavior during meetings and other situations where shared intangible knowledge is exchanged and developed. We can't expect to get a total and complete record of a group's behavior, but we must aspire to comprehensive collection if we are to build comprehensive models. Occasional samples, retrospective accounts, or records of a single type of activity will not be sufficient.

Second, I described a basic framework — the mode of activity — in which to describe collective behavior. Its multilevel structure and constituent categories can provide the high- and intermediate-level symbols for models of collective strategy.

Third, I suggested that we can extend the basic ATN formalism to include rules that identify patterns of behavior across as well as within individual protocol streams. The resulting n-dimensional grammars can function as the analytic components that operate within a distributed systems model of collective strategy. However, to develop the methodology and tools needed to build actual models capable of analyzing and characterizing the work of large, multiteam collaborative projects will require considerable research.

Issues for Research

The long-term research goal identified in this chapter is to develop comprehensive, yet detailed models of collective strategy and to use those models to study collaborative behavior. Here, I will discuss several smaller tasks that continue the line of research outlined at the end of Chapter 6 and that contribute to that goal.

- *Identify specific modes of activity that occur in particular collaborative tasks.*

At the end of chapter 6, I suggested as topics for research identifying the habitual activities engaged by groups and, then, the fine-grained processes that occur within them. Because the mode of activity construct described in this chapter includes both, I suggest further testing of it as a framework in which to describe those activities. The next step would be to identify each mode of activity that occurs in the target group(s) and the constituents of each mode — for example, the goal, constraints, and so on, as well as its multilevel structure. To test the construct, we would need to answer several questions. Does the set of modes account for all mediated cognition cycles observed? Can the behavior of different groups working in different situations be described in terms of a specific set of modes? If the modes framework holds up, then we can use it as a basic architectural construct with which to build models of collective strategy. If it does not, the studies that discredit it may suggest alternative constructs.

- *Build a methodology and a set of tools through which to study collaboration.*

To address the research issues described here and elsewhere in this chapter, we will need better methods and better tools. Observing groups and collecting detailed ethnographic-type notes are difficult and time consuming tasks. Observers need training to help them see what is going on in a group and to control their own preconceived notions. They also need guidelines on issues ranging from ethics to consistent coding. On the other hand, machine-recorded protocols pose different problems. Ideally, commercial developers would include tracking functions in their products that conform to some technical standard, but this will be long coming; when and if it does, it will raise many thorny issues concerning uses of the data. In the meantime, we must look to those building research and experimental collaboration systems for tools of this sort. Regardless of their origin, instrumented systems will produce a flood of data that will have to be stored, managed, and analyzed. Thus, we will need automated analysis tools, such as the grammars described previously, that can integrate and analyze multiple types and instances of protocols. We will also need tools to represent results in visual and dynamic forms that enable human researchers to literally see a group's strategic behavior. Consequently, we should regard methodology, itself, as a legitimate area of research and one that will need its own supporting systems.

- *Build specific models of collaboration.*

The goal of this methodology should be to build detailed, yet comprehensive, models of collaboration for particular kinds of projects, such as group writing and software development. Initially, these models will be quite specific — covering a single project, using a particular set of tools, working within a particular organizational structure and context. With experience, we may be able to extend initial models across different groups, tasks, and conditions. The ultimate goal is to develop a general process model of collaboration; the path we follow to do this should include iterative cycles of specific studies followed by generalization and testing.

- *Use insights derived through these studies to build better support systems, educational curricula, and organizational structures.*

A model of collective strategy should not be an end in itself but, rather, an instrument for developing knowledge about collaborative behavior which can be put to useful purposes. It should lead to better collaboration support systems and to new educational curricula that help groups work together more effectively. If we knew the circumstances under which groups work more effectively, we should be able to adjust organizational structures in specific ways so that they support those circumstances as a matter of policy and intent.

Chapter 8

Collective Awareness and Control

The discussion of collective strategy focused on ways to analyze strategic behavior in groups. Thus, its perspective was external, that of a researcher — or someone else interested for a different reason — observing and characterizing patterns in collaborative activities. In this chapter, the focus shifts from outside to inside.

For an individual to perform a task by following a known strategy involves a substantial degree of self-awareness and self-control. The person must address not only the substantive aspects of the task but also his or her own thinking. For example, the individual must recognize goals, select among options, and test work produced against intentions. These actions and decisions generate a continuous sequence of questions. *How do I get this idea across? Does the sentence sound right? Is it consistent with what I said earlier? Which step in the process should I go to next?* If these questions rise to consciousness, they become part of the thought process that constitutes awareness. But they may not. The individual may shift mental activity in a reflexive, seemingly automatic way as if an underlying question had been raised, but remain unaware of any such question. In either case, the individual experiences a continuum of thought that is both the matrix for strategic decisions and, at the same time, is at least partially comprised of those same decisions.

Can an analogous continuum of thought exist within a collaborative group? If such a continuum exists within individual minds, then it obviously exists within the minds of the group's individual members. But I have a hard time conceiving of a continuum of thought for the group as a whole·

This poses a dilemma. Earlier, I identified as the ideal artifacts that are coherent, consistent, as well as clean and simple in their design. When an objective standard exists, the artifact should also be correct, relative to that standard. For convenience, let me refer to these characteristics when applied to the artifact as *intellectual integrity*. An artifact that has intellectual integrity suggests that the mind that produced it was aware of the whole and successfully related all of the parts to some unifying concept. However, for many collaborative projects, the size and complexity of the task precludes total awareness by any one member. How, then, can groups produce the continuity of thought needed to give their work integrity?

For software development, several approaches have been developed to help teams produce systems that are coherent and internally consistent. One approach is to require groups to go through a prescribed sequence of stages, each tied to a tangible milestone. This may help, but it does not guarantee success. Design documents must be verified against code, and the two kept consistent with one another though the lifetime of the system. In practice, this has been hard to achieve, particularly for large projects of long duration. Because we do not yet have a viable model of the collaborative process and because we have few comprehensive studies of collaborative projects, we have no *principled* basis for believing that one method is better than another or under what conditions a given method will produce reliable results.

One approach that has tried to address the one mind problem is the *chief programmer* method. It views the efforts of a system development team as primarily supporting, and thereby increasing the productivity of, a single highly skilled programmer (Mills, 1968; Baker, 1972), analogous to the team that supports a chief surgeon in the operating room. A chief programmer is responsible for the architecture of the system a team is building, and he or she writes the critical parts of the code. Supporting programmers write individual modules designed by the chief programmer and code the less critical parts of the system. The team also includes an assistant chief programmer and other supporting personnel, such as a librarian and secretary. Because one individual is responsible for the architecture of the system and for overseeing the work of subordinates, the design of the system is thereby held in a single mind — the chief programmer's. The problem with this method is that it does not scale to teams larger than six or eight members or to large projects that require hundreds of programmers and, hence, multiple teams.

Thus, neither ignoring the one mind problem or trying to solve it in a literal manner have proved satisfactory. A different approach is to formulate a concept of *collective awareness* as an abstraction and then try to use that abstraction to find ways of enabling groups to function more coherently and more consistently. One way to do this is to look at the problem from an operational point of view.

Conceptual processing in individual human minds takes place under executive control. A metacognitive process appears to run in human short-term memory, making decisions about intentions and, in turn, activating specific cognitive processes — such as memory accesses, relational operations, and decision-making functions — that carry out those intentions. When a particular functional process completes its task, control returns to the executive process. Of course, the executive process may never be entirely passive, because it can interrupt functional processing when some unusual circumstance occurs that requires immediate attention, like answering the telephone or responding to a fire alarm. The goal, then, is to identify a comparable executive process that operates — or could operate — in groups.[7]

In the remainder of this chapter, I examine issues of awareness and control from a functional point of view. More specifically, I differentiate among several kinds of awareness that exist within collaborative groups in order to identify, albeit in a limited and abstract form, a concept of collective awareness that may help groups achieve a degree of integrity in their work comparable to that sometimes achieved by a single good mind working alone. I then look briefly at issues of collective control, analogous to individual self-control.

---

[7] Metacognitive issues, such as awareness and control, have not received extensive attention from the cognitive science community. One study of note is Klatzky (1984). With respect to distributed cognition, see Rumelhart and McClelland (1986) for a discussion of fine-grained operations, but not large-grain problem-solving and knowledge-construction tasks, as considered here.

Awareness

    Awareness in the sense that is considered here is an analog of self-awareness. Human beings are aware of past experiences — their own but also those of others they learn about and share vicariously — by virtue of their access to their respective long-term memories. Individuals are also aware of their own existence and their own thinking. Two analogous forms of awareness can be identified for collaborative groups: awareness of the group's long-term memory and awareness among the members of one another. Let's begin with the first.

    The collective long-term memory has two parts: the artifact and the body of shared intangible knowledge. In chapter 6, the two were treated separately; with respect to awareness, they should be considered together. Although many of us would admit the possibility that intelligent agents may eventually traverse the artifact with some degree of awareness of its structure and content, for now any such artificial awareness is too limited to be relevant for a concept of collective intelligence. Consequently, I assume that awareness of the artifact will exist only in the minds of the human beings who comprise the group. Thus, awareness of the artifact is part of the group's intangible knowledge.

    When we consider the level of awareness required for a concept of collective awareness, we should not set a goal that is higher than necessary or is higher than that found in individual human beings. None of us has total awareness of our respective long-term memories. At any one moment, we are aware of only a very small part of it — the contents of working memory. Although we may "sense" that we have accessed the very ideas we want for some purpose, we can never be sure that other, still more useful ideas are not stored in our long-term memories but have not been recalled. Thus, we can never be sure that what we are aware of is the most relevant knowledge potentially available to us. Rather, we activate and attend to portions of long-term memory and then move on to other portions. Thus, awareness of a large conceptual structure is a collage of partial awarenesses, generated at different times, at different granularities, and at different levels of abstraction. Consequently, we should not expect collective awareness to be a total and complete awareness of the artifact, at least not an awareness that is activated at any one time.

Awareness in groups exists at several levels of detail. The most general is the body of intangible knowledge that is shared by all members of the group. It includes the overall goals of the project, its ways of operating, the strategies it uses to develop the artifact, its current status and problems, the relation of the project to the external environment, and so on. It also includes that comprehensive overview of the artifact held in common by the group. This awareness is not deep, if the project is large, but it provides each member with a sense of the whole.

At the other extreme is the deep, detailed, often technical, knowledge held by individual members. Depending on the project, a single individual is often responsible for a particular part. That person is expected to have deep, direct knowledge of the corresponding portions of the artifact as well as the underlying issues that inform them. Thus, the person's intangible knowledge goes well beyond what is present in the artifact to include reasons why the construct was built as it was, alternative designs considered and rejected, as well as more general knowledge of the task domain. The person uses this depth of knowledge to produce new components that become part of the artifact. Thus, the level of awareness and expertise required to generate a segment is significantly greater than that required for another person to understand it.

Between these extremes of general, shared knowledge and deep, individual generative knowledge is an intermediate level. It represents a form of awareness that is not often recognized in collaborative groups but may, ultimately, be the most important with respect to the integrity of the group's work. It is *thick* knowledge of adjacent or nearby areas. It takes the form of understanding, rather than generation. Thus, it is shared with the individuals or the team responsible for developing other parts of the artifact, but it is not as deep as their knowledge nor is it shared with the entire project. The most important function of thick knowledge is to inform work in the person's or team's own area by providing a kind of peripheral vision that extends into nearby areas. Thus, it provides a context for the interfaces between areas.

Developing this type of ancillary knowledge is important because no specification — formal or informal — can anticipate all of the issues that will arise in detailed design and implementation. Indeed, a specification must be ambiguous relative to its implementation, otherwise it could be executed directly and its abstract expression would include as much detail as its implementation. Consequently, designers and programmers must interpret specifications, choose among alternative algorithms, make assumptions about processes that operate on the other side of interfaces, and otherwise exercise judgment. Bill Curtis and his colleagues at MCC (Microelectronics and Computer Technology Corporation) discovered that one of the scarcest and most important resources found (or not found) in industrial software projects across a number of different organizations is knowledge of the application area, which informs global design decisions (Curtis, Krasner, & Iscoe, 1988). Such knowledge can be considered a form of thick knowledge that extends across the interface between project and potential users and provides a context for requirements and specifications.

Thick shared knowledge can be developed through informal interactions, such as conversations, but it can also be developed through more formal mechanisms, such as reviews. Let's focus on reviews, because they can be institutionalized more easily than informal contacts. A technique used by many software projects as well as other kinds of projects is a type of formal review called a *structured walkthrough* (Yourdon, 1989). When a segment of a design document or a segment of computer code is available, the individual or team responsible assembles a group of colleagues, distributes copies of the relevant sections, and then during a structured meeting "walks" the group through the document or code. The primary goal is to animate the ideas or content for the portion of the artifact under review in the minds of the participants so that they can point out problems not foreseen by the developers.

Although a walkthrough is a time-consuming process, it requires far less time to review a portion of the artifact than it does to create it. Structured techniques make it possible to review a 20–30 page document or a comparable module of code in an hour or two of meeting time plus a similar amount of preparation time. On the other hand, generating the material to be reviewed is likely to take several orders of magnitude more time. Thus, we can identify a *slow build/fast review* cycle for creating and for comprehending portions of the collective long-term memory — the artifact — that is analogous to the *slow write/fast access* cycle identified by Newell to differentiate between the time required for encoding and storing concepts in human long-term memory and the time required for accessing them.

If we step back and look at the project as a whole, we can see that expanding intangible knowledge through slow build/fast review cycles can help members of a group develop fields of awareness in which their own knowledge is deep and concentrated over a relatively small portion of the artifact but extends outward with less, but still substantial, depth over a much larger area. Their respective fields of knowledge overlap with one another at the edges. These borders of *thick shared knowledge* can help members in adjacent teams keep their respective parts of the artifact consistent with one another. If all the teams or individuals in a project develop this type of expanded peripheral awareness, the group as a whole will have built a segmented, but overlapping awareness of the entire artifact.

This discussion of awareness has necessarily been abstract. Let me end it with a metaphor. My reading lamp is a hanging lamp with a handmade paper shade that resembles a bell jar (a cylinder with a rounded top). It is 14 inches both across the diameter of the cylinder and in height. Irregular pieces of paper, each 8–10 inches across and in muted shades of cream, pink, and tan are glued together to form the shade. The pieces do not abut one another; rather, each overlaps a half-inch or so with its neighbors. The overall shape is smooth and regular, enclosing a recessed bulb that casts light through the open bottom of the cylinder, and the whole thing glows softly.

Think of the individual awarenesses of group members as analogous to the individual pieces of paper in the lampshade. The boundaries of individual knowledge do not abut; rather, they overlap with one another. Thus, they share a common border — a boundary that is not a line, but has significant area. When this is the case, a portion of the artifact developed by one member will be informed by that individual's thick shared knowledge of nearby segments being developed by colleagues, and vice versa. In addition, portions of the artifact larger than a given individual's primary responsibility will have been held in that person's mind. Thus, the one mind condition will apply to a set of overlapping segments of the artifact.

When all the partial thick awarenesses are assembled, they form a whole that encloses the artifact. It is important that the surface of this collage be continuous, so that all parts of the artifact are surrounded. When this is the case, the artifact will have been subjected to the one mind condition, albeit in a piecemeal but continuous fashion.

Thus, we can identify three forms of awareness with respect to a group's long-term memory: close, detailed awareness of particular segments of the artifact; less detailed, but still substantial, awareness of adjacent parts of the artifact; and the much thinner awareness of the artifact as a whole that is shared by the entire group. A different kind of awareness is the awareness members have of one another.

In addition to awareness of our respective long-term memories, we are also aware of our own thought processes. We can recall earlier instances when we thought about a particular problem or issue. At least occasionally, we may observe our thinking as it occurs to see that it is momentarily distorted by emotions or other socially induced factors. We also have a general sense of what we know and do not know, areas in which our knowledge is deep, and those in which it is not. These are different forms of self-awareness in which we briefly seem to step out of ourselves yet observe ourselves as a functioning mental process. Comparable forms of awareness exist within groups. Within the terms of this discussion, we can consider the problem with respect to the collective processor and the awareness one processor has of the other processors and of the system as a whole.

One of the primary reasons for assembling a group is to assemble the expertise required to carry out a project. For complex tasks, not all of the required expertise will be found in one head. Although it is conceivable that the work of the group could be partitioned so that requisite expertise is always matched with assigned task, this is seldom the case. Often an individual must call on his or her colleagues for help. The issue, then, is providing the group as a whole with a collective awareness of its members' respective specialized knowledge and expertise. Some groups refer to an individual with specialized knowledge as a *guru* in that area. Thus, an extremely valuable resource for a group is shared knowledge of who is a guru on what. Some computer systems, such as UNIX, even have built-in facilities for recording and accessing information on gurus. But, regardless of the mechanism, a group's knowledge of its gurus is comparable to an individual's knowledge of his or her own depth of knowledge in specific areas.

Another form of awareness is the awareness at any given moment one processor has of the other processors. For example, one member of the group may be aware, or wish to know, that another member is working in a nearby part of the artifact. This behavior is monitored at a very low level by the collaboration support system in its concurrency control mechanisms to insure that two members do not try to change the very same part of the artifact at the same time. However, these mechanisms do not prevent one member's access from blocking that of another or of one member's subsequent work affecting earlier work done by another. Higher level interactions such as these must be permitted, but groups may also need help in monitoring domains of activity. For example, members may want tools that can provide a visual image of the artifact and show them where they are working within it. They may also want to see where colleagues are working. They may even wish to see a display over time of the "tracks" left by colleagues.

A third form of awareness involves the interaction between social and intellectual processes operating within the group. I discussed the fine-grained part of this issue in chapter 6 with respect to group-mediated cognition cycles. There, social and conceptual actions were seen to interleave as groups construct and/or use shared knowledge. A set of larger-grained issues has to do with the more overt effects members of the group have on one another and on their collective work. It would be nice if groups were purely intellectual organisms. But they are not. Tensions exist; friction occurs. These developments are inevitable. For the most part, they remain at the level of distraction, but they can become more intense and affect conceptual work. For example, one member may oppose an idea voiced by another not because the idea is bad but because of who said it. The opposite condition — supporting an idea because of friendship or attraction — is equally bad. These so very human situations are unlikely to go away, but a group should be aware of them and through its control and decision making procedures, try to insure that the integrity of its work is not compromised by them.

Thus, a group should be aware of itself as a dynamic, functioning organism as well as be aware of the artifact it is developing. From an experiential point of view, I still cannot envision a single, integrated awareness for a collective intelligence, but I can imagine it as a structural entity in the form of a collage of partial, but overlapping partial awarenesses. I can imagine how this form of collective awareness might function. And I can imagine how we might develop methods and tools to support it and to help groups develop it.

At the end of chapter 4, I briefly discussed an objection raised by Newell to the concept of collective intelligence, based on the limited rate at which human beings can transfer knowledge from one to another. Newell asserted that for a group to behave as a coherent rational agent, each member of the group must know everything all of the other members of the group know. He is no doubt right that this is an impossible condition, but it may be too strong a requirement. Although not the complete knowledge Newell calls for, this composite of deep, comprehensive, connected awarenesses that comprise what I have called collective awareness may be sufficient to produce artifacts that are coherent and internally consistent to a degree comparable to those produced by a single good mind. If this is the case, then we can say that for practical purposes, the group has achieved a form of collective intelligence and thereby met Newell's objection.

Control

Control within groups has been studied from a variety of perspectives, including organizational theory, interpersonal relations, the characteristics of effective leaders, the impact of technology, patterns of communication, and the dynamics of groups over time. Much of that research focuses on social, as opposed to intellectual, factors. In chapter 1, I constrained this discussion to intellectual tasks. Thus, although the social dimensions of collaboration are important, for the kinds of groups being considered here, intellectual behavior is fundamental. Consequently, in this chapter I will consider control from a perspective in which conceptual and social dimensions are merged.

The form of control that is discussed is an extension of *self-control*. Self-control is an executive, metacognitive function that monitors the behavior of an individual — both mental and physical — and adjusts it in accord with some structure of goals, self-image, and/or set of external conditions. Thus, it monitors and responds to mental processes that lie below consciousness, to outside stimuli, and to properties and processes in the physical body. Although it is closely related to rational processing, it is not, itself, entirely rational.

I refer to the analogous form of control within a group as *collective control*, consistent with the terminology used in earlier discussions for other parts of a collective intelligence; however, I do not mean to imply that this form of control is necessarily democratic.

Collective control includes two large components: an organizational component and an intellectual component.

The organizational component is concerned with the group's operation and the procedures it uses. It is concerned with establishing an overall strategy for the group, setting priorities and goals, monitoring progress, and resolving organizational conflicts within the group. It is also responsible for obtaining the resources needed by the group and for interacting with the outside world. While essential, the organizational control function does not directly engage the substance of the group's work or directly manipulate the artifact. Consequently, it is similar to the executive function that monitors and controls what Vygotsky referred to as "lower" mental and physical processes.

The intellectual component does directly engage conceptual substance, because it is concerned with building a coherent, consistent structure of ideas. Consequently, this form of control places a high priority on "getting it right," recognizing validity as a primary requirement for the group's work. It may also try to achieve intellectual elegance — not as an end in itself, although clean, simple conceptual structures are often compelling, but because work with these characteristics is easier to understand, to communicate, and to maintain. This form of control includes "higher" mental functions, as Vygotsky used the term, including establishing a basic set of terms and concepts, constructing an overarching conceptual framework, and expanding and implementing that construct. Throughout this process, the control process monitors conceptual construction in order to modify the overall artifact design and to reconcile inconsistencies and differences of opinion that arise. As a result, it is responsible for the evolution and integrity of the artifact.

Collective control is, thus, the union of two types of executive functions — one organizational, the other intellectual. It is an abstraction that becomes actual in both the formal and the de facto control structures that exist within groups and in the individuals who function as leaders and/or occupy positions of authority. Thus, collective control can be achieved by a number of different organizational structures and styles of leadership. My own experience suggests that although a strong leader invested with authority is required, most often he or she leads best who leads least. If we tap on this simple maxim, it unfolds into the much larger concept of collective control I am describing.

First, intellectual integrity is best achieved if all members of a group *try* to achieve it in their individual work and in the areas where they have awareness and responsibility. In this way, they care what happens. Thus, developing a sense of shared ownership and responsibility throughout the group is important.

Second, all members must feel they can speak freely on substantive issues. Members doing detailed, technical work are often in the best positions to monitor and report problems. Those with leadership responsibilities, at all levels, must listen and respond. Otherwise, information will not flow freely, and the work of the group will be "brittle." The Challenger and Three Mile Island disasters were both dramatic failures caused by brittleness, in the sense that crucial information either did not flow or was not attended to across boundaries in the system. Less dramatic failures occur all the time when groups fail to achieve collective self-control.

Third, those who function as leaders perform acts of selection as often, if not more so, as they perform acts of generation. People in positions of authority do not have a franchise on good ideas; in fact, just as those directly responsible for detailed substantive work are in the best position to see problems, they are also the ones most likely to see new possibilities. However, those in leadership roles are often in a better position to select among new ideas, including testing them against the overall structure of the artifact and working out inconsistencies caused by adopting them. As the conceptual structure evolves, they are also responsible for articulating a new view of the whole to update shared intangible knowledge in the group.

Thus, the role I have sketched for a group's leaders is more a matter of perspective than authority, although authority must ultimately be vested in those leaders. In an organizational structure that has line authority, a team leader who works with several individuals is in a position to look over all of their shoulders and see how an idea generated in one context affects other contexts. It is conceivable that this same function of selection and reconciliation could be done by the team, itself, operating as a committee of the whole and without a designated leader. They could use their collective experience and knowledge to evaluate new ideas generated by any one of them and collectively make decisions that affect their part of the artifact.

Where this form of control breaks down is when there is disagreement in the group. In those instances, making a decision by vote in a project that is primarily intellectual is not an acceptable solution; the majority may simply be wrong! The reason for this is the fact that the artifact is a material object that the group is responsible for building in such a way that it is coherent, consistent, and correct. Decisions that invalidate the integrity of the artifact are wrong, regardless of how they were arrived at and regardless of how many people agree with them. When consensus does not exist, some *one* individual must decide which option is (most) consistent with maintaining the intellectual integrity of the whole. Such a decision must be made on substantive grounds, not on the basis of its effects on social or organizational concerns.

Thus, the work of the group *may* happen by consensus; it probably *will* happen by consensus most of the time; but when consensus does not exist, some individual must have the authority to step in and make the decision that tries to maintain the intellectual integrity of the group's work, within the limits of that individual's capabilities. Thus, although I can admit ad hoc and network-based organizational structures that function most of the time across most of the group's activities, I cannot envision a group structure that can *reliably* produce work that is coherent, consistent, and, possibly, elegant that is not, ultimately, hierarchical, in the sense described here.

Finally, let me point out that collective control resembles, but is not identical to, the chief programmer model. It tries to achieve integrity and, perhaps, elegance in collaborative work by having the entire design or conceptual structure come together in a single mind — that of the overall project leader — but at a (possibly high) abstract level. However, it differs from the chief programmer model in two important respects. First, it assigns the leader a role that is much more integrative, based on selection and comprehension rather than on generation. By contrast, the chief programmer is expected to generate the primary architecture and important portions of the code. Second, it assumes that there may be other analogous leaders that serve similar roles with regard to individual teams. Consequently, the model can be scaled by including intermediate levels, consisting of groups of teams. If the project is large and includes multiple groups and levels, thick shared knowledge should overlap vertically between levels — just like it overlaps horizontally between teams on the same level — in order to provide continuity over the entire project. Thus, collective control, and the forms of project organization and integrated behavior it implies, achieves many of the benefits of the chief programmer model, without several important limitations.

In this chapter, I have looked at analogs for two of the more complex and elusive metacognitive functions that permeate human intelligence.   The first is awareness.   When we identify the characteristics of human intellectual work that we value most highly — coherence, consistency, correctness, and, elegance — it is difficult to imagine how work with these attributes could be produced without that structure of ideas having been held in its entirety by a single mind, if not actually produced by that mind.   However, by considering awareness from a functional point of view, we may be able to construct mechanisms that can enable groups to achieve comparable results.   One is a collage of partial but overlapping awarenesses based on thick, shared knowledge distributed over the group.   Another is awareness within the group of the varied expertise held by its members.

The second issue considered was control.   A mind capable of producing large artifacts that have intellectual integrity must also be disciplined.   Not necessarily in a rigid way, but with enough self-control, informed by self-awareness, that it can test the constructs it produces against one another, against more general principles, and, perhaps, against some deeper aesthetics it has come to associate with "getting it right."   The analog for self-control in a collaborative group is a collective control that balances hierarchical authority, required to resolve conflicts, with mechanisms that distribute responsibility throughout the group and generate vested concern for the integrity of their collective work.

Although it remains difficult to imagine how groups can achieve the same coherence and the same grace in their work that is sometimes achieved by individual minds working alone, we do not always have that option.   Consequently, we must try to formulate mechanisms, such as those discussed here, that approximate the same functional characteristics within groups.

Issues for Research

A number of research issues emerge from concern for collective awareness and control.   In the long term, we will need new tools and new methods for studying collaborative behavior, such as those described in chapter 7, to make these problems tractable. However, it is not too early to begin addressing questions such as the following, because even partial results would yield significant benefits.

- *How do groups develop awareness of expertise distributed within the group and, in turn, use it effectively?*

Because most collaborative projects include individuals with complementary knowledge and skills, knowledge about who knows what must be distributed throughout the group. Also important is helping team members make good decisions about when to dig for knowledge on their own — from books, databases, etc. — and when to seek help from colleagues. A related problem is communication across knowledge boundaries. An individual seeking help may have difficulty finding a person who has the information he or she needs if the person does not know the terms in which to express that need so that a person who has the knowledge will recognize it as related to his or her expertise. Research in AI, library science, and automatic translation could be applied to this problem.

- *What are the properties and uses of thick shared knowledge?*

How is thick knowledge, developed by individuals with respect to adjacent areas of a project, built? How is it used? How much is needed? What happens if it is too thick or too thin? Can its development be institutionalized or is it a matter of individual choice and behavior? In what ways does thick "vertical" knowledge differ from thick "horizontal" knowledge?

- *How can we identify, analyze, and represent the collage of knowledge that surrounds the artifact?*

If a large project is surrounded, first, by a thin membrane of knowledge shared by the group as a whole and, second, by thick patches of more specialized knowledge, how can we characterize these bodies of knowledge? What specific concepts and structures do they consist of? How do the pieces fit together? How much should they overlap? Is it important for groups to share certain kinds of knowledge but not others? How can we tell if the artifact is completely surrounded? Does it matter?

- *What is the relationship between organizational structure — de facto as well as defined — and intellectual integrity in the artifact?*

If we characterize the relationships and structures that comprise collective control, can we then see how it affects the integrity of the group's work? Do properties in the artifacts correlate with specific patterns of behavior? If so, can we trace those patterns to particular control structures that generate or influence them? Do some organizational structures and procedures result in more cohesive work than others?

- *Which traits or experiences enable a leader to work effectively with structures of abstract symbols?*

Why can some people work with highly abstract symbols better than others? For large conceptual constructs, the architecture is formed at a level of high abstraction. Thus, each symbol or element in the design stands for a much larger component that, itself, may be quite deep. Consequently, high level symbols have long "tendrils" attached to them. If the design is to work well, those tendrils must descend gracefully and they must not get tangled with one another. Some individuals seem to have an intuitive feel for what lies beneath the symbols they work with, even when they have little direct knowledge of details at lower levels. They use that feel to produce designs that make clean separations that work well all the way down, or they use it to point out problems of decomposition in the designs of others. Is the skilled designer or project leader who works at a high level of abstraction necessarily someone who has worked his or her way up through the ranks and thereby developed this feel? Or, is it a native characteristic? Can it be learned? Can it be taught, either through training or mentor relationships?

- *What makes a conceptual structure elegant?*

What, exactly, does it mean to say that an intellectual product is simple and elegant? Is it only something we recognize when we see it? Or, can we develop parameters that will give us a more analytic sense of the characteristics that underlie such products? Is clean design related in some inherent way with the content domain? That is, are "seams" fundamental to a domain and, hence, discovered by a designer, or are they ultimately arbitrary and, hence, constructed?

How far down should an elegant design be expected to extend before it dissolves into arbitrary and/or messy detail?

- *What enables a group to produce elegant work?*

If it's easy for groups to design camels, why can't they learn to design eagles? Assuming we can arrive at a more basic understanding of what constitutes elegant design, how can groups achieve it? Is it a function of the individuals that comprise the group, its leader(s), the tools it works with, its procedures and organizational structure, and/or the environment in which it is located? Can groups learn this skill? If so, can we develop curricula and instructional programs to help groups develop it? One way to pursue this would be through case studies. If we can identify groups that have produced elegant products in the past, perhaps we could retrace their steps to see what enabled them to do so. Best of all would be to follow work in progress that turned out this way. What signals might alert us to such a group?

Chapter 9

Conclusion

The primary goal for this book was to help move discussion from a vague *notion* of collective intelligence to a reasonably well-defined *concept*. The issue was constrained to groups of limited size and duration doing intellectual tasks. By focusing on conceptual work, I was able to consider collaborative groups as a form of information processing system, analogous to Newell's and Simon's IPS model of individual cognition. The bulk of the discussion, then, became an examination of that system and its main components.

A collective memory system was described that includes subsystems that provide a collective long-term memory for tangible knowledge, built and maintained in a computer system, and for intangible knowledge, carried in the heads of the human beings that comprise the group. The memory system also includes working memory for both types of information.

A collective processor was described next. It includes the fine-grain operations used by groups to develop, access, and maintain the information stored in the memory system. Three basic types of processes were discussed: computer-, group-, and conference-mediated cognition cycles, depending on the situation in which they occur and the technology being used. The collective processor as a whole was described as a loosely coupled distributed system that includes multiple independent processors, joined by communications and social networks.

Collective strategy enables coherence in collaborative work. Individual processes occur not in isolation but in purposeful sequences. We can think of these strings of operations as being similar to statements in a language intended to accomplish a goal or to communicate a message. The system responsible for generating sequences of operations is analogous to the grammar individuals use to generate strings of words. Consequently, one way to study strategic behavior is to identify the rules for a grammar that can parse the sequences of operations observed in groups. Such a grammar would represent an analytic model of collective strategy.

The discussion concluded by considering two metacognitive issues: collective awareness and collective control. Many projects are too large and too complex to be understood by any one person. Yet, we would like to see groups produce work with the same integrity and consistency sometimes found in work produced by a single good mind working alone. By developing thick, overlapping areas of shared knowledge, perhaps through slow build/fast review cycles, groups may be able to piece together a form of collective, but distributed awareness that is sufficiently coherent to achieve this goal. Control must also be distributed over a group. Otherwise, information will not flow across boundaries, and the group and its work will be brittle. However, although many decisions can, and probably should, be made by consensus, authority must ultimately be centralized in order to resolve disagreements and to preserve the integrity of the group's work.

What I hope has emerged from this discussion is an *image* of collective intelligence. It is a highly abstract image that references other systems and models as well as the physical world. However, I have tried to sketch it in enough detail so that the reader can "see" it in the mind's eye — both the shape of the whole and its main components. I have also tried to describe it "in motion" so that it can be seen as a dynamic system, in the process of building large conceptual structures. But, it is only a sketch — more suggestive than definitive. A great deal of research will be needed to flesh out its outline and to correct its mistakes. Nevertheless, I hope the discussion, even in its current form, has shown that a theory of collective intelligence is possible.

Perhaps that work can be motivated by considering the potential value a theory of collaboration could have. In the preface, I identified two primary audiences for this book: people involved in research in collaboration behavior and systems, and, second, people who work collaboratively or are simply interested in the topic. Let's look at this issue for both groups.

Currently, research in Computer Supported Cooperative Work is highly fragmented. Most teams focus on only a small part of the problem with little thought about how their work fits into a larger whole or how contextual factors impact their results. Yet, patterns of behavior observed in one situation may not appear under other circumstances. For example, behavior in an early design meeting held in a group's normal working environment may be quite different from behavior that occurs in a laboratory setting or in a down stream review session. The same may be true for groups working on small, contrived problems versus large, actual problems in which they have a vested interest. If the research community saw itself as working toward a general theory of collaboration, individual projects would be encouraged to think more about how their part of the problem relates to work going on elsewhere. And, as that theory emerged, it would provide continuity across disparate studies.

This problem of isolation in the CSCW research community is not more pronounced than in other areas, but it is more ironic. Indeed, if the topic of research is *cooperation*, should not those in the field be working cooperatively, themselves? Working on a general theory of collaboration is tantamount to working within a paradigm. But, as Kuhn (1962) observed, the emergence of a paradigm marks the mature stage of a discipline. We may not be at that point yet in collaboration research. Indeed, not everyone believes a theory of collective intelligence is even possible. But, regardless of one's position on that issue, it is not too early to identify a common set dimensions that encompass all research in this field. By identifying those dimensions, and the points that fall along them, we can begin to see where individual projects lie, how research done by one group relates to that of another, and ways groups could work together more effectively.

To illustrate this idea of research dimensions, consider the five types of information built and used by collaborative groups: target and instrumental products, shared and private intangible knowledge, and ephemeral products. Each of these five types can be associated with a point along an *information type* dimension.[8] A number of studies — previous or potential — can then be viewed as examining movement of information along this dimension. For example, considering how intangible knowledge becomes part of the artifact can be viewed as a study of the transformation processes that map types found at one point to types associated with a different point. A related question is how the artifact, which represents general tangible knowledge with respect to the group, is comprehended by a new member and, thus,

---

[8]   This dimension is *ordinal*, consisting of identifiable states and implying no concept of distance between points. Other dimensions are *scalar*, implying a measurable distance relationship between points.

transformed into private intangible knowledge relative to that person. Similarly, one could consider the roles ephemeral products play in both of these processes.

Other dimensions include the spectrum of tools groups use, the duration over which behavior is observed and characterized, physical space, the size and coherence of the group, and the various processes groups use to do their work. Possible points along these dimensions include the following:

Information   Type

    intangible private
    intangible shared
    ephemeral
    tangible instrumental
    tangible target

Tools

    noncomputer
    computer applications
    database and/or distributed file system
    hypermedia
    data communication
    computer conferencing
    audio/visual communication
    transparent wide area network
    intelligent agents
    observational

Time

    10 sec
    100 sec (several minutes)
    1k sec (~15 minutes)
    10k sec (several hours)
    100k sec (several days)
    1m sec (several weeks)
    10m sec (several months)
    100m sec (several years)
    1b sec (life's work)

Space

> office
> cluster
> floor
> building
> site
> 1-hour travel
> 1-day travel

People

> individuals
> informal coalitions
> teams
> collection of teams
> group as a whole

Processes

> cognitive/conceptual
> metacognitive
> social interaction
> mediated
> organizational
> collective/distributed
> algorithmic

Dimensions become more interesting when they are considered relative to one another, like x,y axis. Two dimensions can define a plane of points, whereas three or more can define spaces. Each point in one of these planes or spaces can be associated with a particular set of issues. For example, we could ask which tools and functions provided by a collaboration support system are used to work with which particular types of information? This question involves two dimensions: *tools* and *information types*. Each point represents a particular combination of tool and information type, and a value for a given point might indicate whether the particular combination was observed in a given group. Similarly, considering whether particular tools are more or less useful for groups located in the same cluster of offices versus a widely distributed group involves the *tools* and *space* dimensions. If a study further broke the question down by type of information, it would involve all three dimensions. We could continue expanding the issue by asking whether the effects are found at all stages of a project or only at certain times; whether the distribution of processes in groups differ according to the size and organization of the group; and so on. Thus, dimensions can be used individually, in pairs, and in higher order combinations.

If researchers could agree on a common set of dimensions, projects could use them to suggest ways their respective agendas might be extended incrementally to make their research more comprehensive and to identify other groups with whom they might collaborate or share data. Conversely, a common set of dimensions could also be used to suggest ways large studies might be divided among several research groups.

An agreed upon set of dimensions could also be helpful for developers of collaboration systems. Most current systems support a limited range of activities, such as shared asynchronous work (e.g., Lotus Notes) or group editing (e.g., Aspects). No commercial system supports the full range of tasks groups carry out, and I am aware of only one research system other than ABC that supports both synchronous and asynchronous work (i.e., SEPIA). Consequently, switching between individual and collective work is cognitively disruptive and requires moving from one computing context to another, with no guarantee that data built in one can be imported into the other. However, if developers viewed requirements against a backdrop of potential functions, the specific tasks those functions support, and the varied conditions under which groups work, it could help them see ways in which their systems could be extended to make them more useful or easier to use. By identifying major seams in the design space, a common set of dimensions could also facilitate developing standards for collaboration systems. Standards, in turn, could have a major impact on future designs, enabling data to be passed easily across boundaries and leading to more modular software that could be used in combination.

Thus, a long-term commitment to building a theory of collective intelligence and to building a more structured awareness of its associated research space could help the CSCW community work with greater continuity and coherence as well as inform system development.

A theory of collective intelligence could also benefit people who work collaboratively but are not researchers or system builders. I have already mentioned one potential benefit — better support tools. Other benefits are related to the ways in which a theory could help groups work more productively.

The issue boils down to the proposition that if the people who form groups had a better understanding of the collaborative process, they would be able to work together more effectively. That understanding would include a mental image of the overall process, strategies to guide them through it, and a view of their own roles in that process, including relationships with other individuals and/or subgroups. It should also help them see, at any given moment, where a project is relative to its long-term agenda and schedule. This proposition assumes that knowledge and awareness will result in better performance, both by individuals and the group as a whole. This may or may not be true, but it is an hypothesis worth testing. In the meantime, I would argue the point by analogy.

A large part of learning any knowledge-construction skill is learning strategies, particularly after one has mastered the rudiments of the skill. Implicit in a strategy is a comprehensive view of the task. I suppose one could argue that specific strategies can be regarded as individual large-grain processes to be used in particular circumstances without the individual having to think about the overall task, but that seems a perverse point of view. Instead, the working assumption in most skill development training seems to be that people need an overview of the process in addition to specific operations they can perform. I am simply arguing that groups need the same thing with respect to collaboration.

We should note, however, that collaboration, as a skill, differs from most conventional skills. It is inherently a second order phenomenon. That is, it is a metaskill that includes first order skills as components. For example, when one takes training in writing, one does so in order to write better documents or to write them more easily. Thus, there is a direct relationship between the skill — writing — and the results of applying that skill — a document. Collaboration is different. It is a layer of behavior that lies on top of conventional skills. For example, groups frequently co-author documents. Writing is the primary skill that actually produces the document; collaboration is the secondary skill — important, but instrumental.

This is as it should be. Nevertheless, it would be a healthy development if a theory of collective intelligence could lead to pedagogy for developing collaborative skills. Students at all levels could be taught collaboration skill, in addition to learning basic, first order skills. A collaboration curriculum could include instruction and practice in specific modes of activity. Thus, groups could go through explicit brainstorming activities, learn how to give and receive oral briefings, perform structured reviews of one another's work, do cross editing on co-authored documents to produce a consistent tone and voice, and so on. They could also learn strategies for using these various modes. And they could learn how to make effective use of both the computer and conventional tools available to them. Although collaboration training could be tied to specific content domains, such as writing or programming, it could also be taught as an abstract skill in which the specific task or content area is left up to the individual team.

Training in collaboration will, no doubt, encourage that way of working. When this happens, it is likely to be more a response to changing conditions than a driving force. Collaboration is already the predominant mode of work in many industries and in many cultures. That trend will probably accelerate. Although this may lead to greater productivity, it may also challenge deeply held values. In some societies, individuality has been a cornerstone of their cultural heritages. They honor rugged individuals, tell stories about their deeds, and train people to work alone. Consequently, much of an individual's self-image and self-worth may be tied up in his or her sense of uniqueness.

At least one response to this (potential) problem is suggested by the concept of collective intelligence, itself. Individual conceptualization lies at the heart of group-mediated cognition. Indeed, we saw in chapter 7 that group-mediated cognition is driven by the insights of individuals that are then shared with the group. But groups must also make decisions, build substantial bodies of shared knowledge, and work from consensus. And the individuals in a group must know when it is time to exercise free and unrestrained individuality and when it is time to close ranks and move on. Collaboration is, thus, a rhythmic process that oscillates between individual and collective behavior, individual and collective thinking. To be sure, it constrains individuality in some circumstances, but behavior has always been constrained, whether the individual realizes it or not, by the culture, by standards of acceptable behavior, and by laws. But, along with the constraint posed by collaboration comes a compensating benefit — the sense of being part of a group and contributing toward a common goal. Thus, all things in their season. A theory of collective intelligence can help people strike the right balance between individual, independent thought and deed and collective, interdependent forms of the same.

As we look toward the future, the technologies that are emerging promise collaboration and cooperative interaction on a worldwide scale. That much seems certain. What is less clear is to what purposes we will put that technology. No doubt part of it will be devoted to popular entertainment — the proverbial 500 channels of television — and to chat groups — the equivalent of international CB radio. I see nothing inherently wrong with such uses. But we should also use these resources in our work and in our efforts to solve some of the many problems the world faces. It is to those efforts that a theory of collaboration can contribute most. Thus, developing a theory of collective intelligence is more than a matter of casual or academic interest. It could mean the difference between being run over by the technology and harnessing it for worthwhile purposes. It is a goal worth pursuing.

References

Abdel-Wahab, H. M., Guan, S. -U., & Nievergelt, J. (1988). Shared workspaces for group collaboration: An experiment using Internet and UNIX interprocess communications. *IEEE Communications, 26*(11) 10–16.

Abel, M. J. (1990). Experiences in an exploratory distributed organization. In J. Galegher, R. E. Kraut, & C. Egido (Eds.), *Intellectual teamwork: Social and technological foundations of cooperative work* (pp. 489–510). Hillsdale, NJ: Lawrence Erlbaum Associates.

Anderson, J. R. (1983). *The architecture of cognition.* Cambridge, MA: Harvard University Press.

Anderson, J. R. (1990). *The adaptive character of thought.* Hillsdale, NJ: Lawrence Erlbaum Associates.

Anderson, J. R., & Bower, G. (1973). *Human associative memory.* Washington, DC: Winston.

Baker, F. T. (1972). Chief programmer team management of production programs. *IBM Systems Journal, 11*(1) 56–73.

Bellcore, (1989). *The VideoWindow teleconferencing service model* (Special Rep. No. SR–ARH–001424). Morristown, NJ: Bell Communications Research.

Bikson, T. K., & Eveland, J. D. (1990). The interplay of work group structures and computer support. In J. Galegher, R. E. Kraut, & C. Egido (Eds.), *Intellectual teamwork: Social and technological foundations of cooperative work* (pp. 245–290). Hillsdale, NJ: Lawrence Erlbaum Associates.

Birman, K. P. (1989). How robust are distributed systems? In S. J. Mullender (Ed.), *Distributed Systems* (pp. 441–452). New York: ACM Press.

Blakeley, K. D. (1990). *The application of modes of activity to group meetings: A case study* (Tech. Rep. No. TR90–045). Chapel Hill, NC: UNC Department of Computer Science.

Boehm, B. W. (1988). A spiral model of software development and enhancement. *Computer, 21*(5) 61–72.

Brooks, F. P., Jr. (1975). *The mythical man-month.* Reading, MA: Addison-Wesley.

Bush, V. (1945). As we may think. *Atlantic Monthly, 176*(1) 101–108.

Cain, C.; & Reeves, J. R. (1993). *Ethnographic research: Process and methods* (Tech. Rep. No. TR93–046). Chapel Hill, NC: UNC Department of Computer Science.

Card, S. K., Moran, T. P., & Newell, A. (1983). *The psychology of human-computer interaction.* Hillsdale, NJ: Lawrence Erlbaum Associates.

Cook, P., Ellis, C., Graf, M., Rein, G., & Smith, T. (1987). Project Nick: Meetings augmentation and analysis. *ACM Transactions on Office Information Systems, 5*(2) 132–146.

Coulouris, G. F., & Dollimore, J. (1988). *Distributed systems: Concepts and design.* Reading, MA: Addison-Wesley.

Crowley, T., Milazzo, P., Baker, E., Forsdick, H., & Tomlinson, R. (1990). MMConf: An infrastructure for building shared multimedia applications. *CSCW '90 Proceedings* (pp. 329–355). New York: ACM Press.

Curtis, B., Krasner, H., & Iscoe, N. (1988). A field study of the software design process for large systems. *Communications of the ACM, 31*(11) 1268–1287.

Department of Health and Human Services. (1990). *Understanding our genetic inheritance. The U.S. Human Genome Project* (DHHS Publication No. A05). Washington, DC: U. S. Government Printing Office.

Dewan, P., & Choudhary, R. (1991). Primitives for programming multi-user interfaces. *Proceedings of the Fourth ACM SIGGRAPH Symposium on User Interface Software and Technology* (pp. 69–78). New York: ACM Press.

Digital Equipment Corporation. (1985). *VMS system software handbook.* Maynard, MA: Digital Equipment Corporation.

Elrod, S., Bruce, R., Gold, R., Goldberg, D., Halasz, F., Janssen, W., Lee, D., McCall, K., Pedersen, E., Pier, K., Tang, J., & Welch, B. (1992). Liveboard: A large interactive display supporting group meetings, presentations and remote collaboration. *CHI '92 Conference Proceedings* (pp. 599–607). New York: ACM Press.

Engelbart, D. C., & English, W. K. (1968). A research center for augmenting human intellect. *Proceedings of the Fall Joint Computer Conference* (pp. 395–410). Reston, VA: AFIPS Press.

Engelbart, D. C., Watson, R. W., & Norton, J. C. (1973). The augmented knowledge workshop. *AFIPS Conference Proceedings* (pp. 9–21). Montvale, NJ: AFIPS Press.

Ensor, J. R., Ahuja, S. R., Horn, D. N., & Lucco, S. E. (1988, March). The rapport multimedia conferencing system—a software overview. *Proceeding of the Second IEEE Conference on Computer Workstations* (pp. 52-58). Washington, DC: IEEE Press.

Ericsson, K. A., & Simon, A. S. (1980). Verbal reports as data. *Psychological Review, 87,* 215–251.

Finholt, T., Sproull, S., & Kiesler, S. (1990). Communication and performance in ad hoc task groups. In J. Galegher, R. E. Kraut, & C. Egido (Eds.), *Intellectual teamwork: Social and technological foundations of cooperative work* (pp. 291–325). Hillsdale, NJ: Lawrence Erlbaum Associates.

Fish, R. S. (1989). CRUISER: A multimedia system for social browsing. *SIGGRAPH Video Review* (video cassette), *45*(6).

Fish, R. S., Kraut, R. E., & Chalfonte, B. L. (1990). The VideoWindow system in informal communications. *CSCW '90 Proceedings* (pp. 1–11). New York: ACM Press.

Fish, R. S., Kraut, R. E., Root, R. E., & Rice, R. E. (1992). Evaluating video as a technology for informal communication. *CHI '92 Conference Proceedings* (pp. 37–47). New York: ACM Press.

Flower, L. S., & Hayes, J. R. (1984). Images, plans, and prose: The representation of meaning in writing. *Written Communication, 1,* 120–160.

Gaver, W., Moran, T., Maclean, A., Lovstrand, L., Dourish, P., Carter, K., & Buxton, W. (1992). Realizing a video environment: EuroPARC's Rave system. *CHI '92 Conference Proceedings* (pp. 27–35). New York: ACM Press.

Gray, W. D., John, B. E, & Atwood, M. E. (1992). The precis of Project Ernestine or an overview of a validation of GOMS. *CHI '92 Conference Proceedings* (pp. 307–312). New York: ACM Press.

Group Technologies, Inc. (1990). *Aspects user manual.* Arlington, VA: Authors.

Hayes, J. R., & Flower, L. S. (1980). Identifying the organization of writing processes. In L. W. Gregg & E. R. Steinberg (Eds.), *Cognitive processes in writing* (pp. 3–30). Hillsdale, NJ: Lawrence Erlbaum Associates.

Hayes, J. R., & Flower, L. S. (1986). Writing research and the writer. *American Psychologist, 41,* 1106–1113.

Herdan, G. (1960). *Type-token mathematics.* The Hague: Mouton.

Holland, D., Reeves, J., & Larme, A. (1992). *The constitution of intellectual work by programming teams* (Tech. Rep. No. TR92–013). Chapel Hill, NC: UNC Department of Computer Science.

Holland, D., & Valsiner, J. (1987). Symbols, cognition, and Vygotsky's developmental psychology. *Ethos, 16*(3) 247–272.

Ishii, H., Kobayashi, M., & Grudin, J. (1992). Integration of inter-personal space and shared workspace: ClearBoard design and experiments. *CSCW '92 Conference Proceedings* (pp. 33–42). New York: ACM Press.

Jeffay, K., Lin, J. K., Menges, J., Smith, F. D., & Smith, J. B. (1992). Architecture of the Artifact-Based Collaboration System Matrix. *Proceedings of CSCW '92* (pp. 195–202). New York, ACM Press.

Jeffay, K., Stone, D. L., & Smith, F. D. (1992). Kernel support for live digital audio and video. *Computer Communications, 15*(6) 388–395.

Jeffay, K., Stone, D. L., & Smith, F. D. (in press). Transport and display mechanisms for multimedia conferencing across packet-switched networks. *Computer Networks and ISDN Systems.*

John, B. E., & Vera, A. H. (1992). A GOMS analysis of a graphic, machine-paced highly interactive task. *CHI '92 Conference Proceedings* (pp. 251–258). New York: ACM Press.

Klatzky, R. L. (1984). *Memory and awareness: An information-processing perspective.* New York: Freeman.

Krol, E. (1992). *The whole Internet: User's guide & catalog.* Sebastopol, CA: O'Reilly.

Kuhn, T. S. (1962). *The structure of scientific revolutions.* Chicago: University of Chicago Press.

Kupstas, E. (1993). *Patterns of interaction in same-time, same-place collaborative programming* (Tech. Rep. No. TR93–006). Chapel Hill, NC: UNC Department of Computer Science.

Lansman, M. (1991). *Organize first or write first? A comparison of alternative writing strategies* (Tech. Rep. No. TR91–014). Chapel Hill, NC: UNC Department of Computer Science.

Lansman, M., & Smith, J. B. (1993). Using the Writing Environment to study writers' strategies. *Computer and Composition 10*(2) 71–92.

Latour, B., & Woolgar, S. (1979). *Laboratory life: The social construction of scientific facts.* Beverly Hills, CA: Sage.

Leland, M. D. P., Fish, R. S., & Kraut, R. E. (1988). Collaborative document production using Quilt. *CSCW '88 Conference Proceedings* (pp. 206–215). New York: ACM Press.

Lotus Development Corp. (1993). *Lotus Notes administrator's guide.* Cambridge, MA: Lotus Development Corp.

Mackay, W. E., Guindon, R., Mantei, M., Suchman, L., & Tatar, D. G. (1988). Video: Data for studying human-computer interaction. *CHI '88 Conference Proceedings* (pp. 133–137). New York: ACM Press.

Malone, T. W., Lai, K. Y., & Fry, C. (1992). Experiments with Oval: A radically tailorable tool for cooperative work. *CSCW '92 Conference Proceedings* (pp. 289–297). New York: ACM Press.

Mantei, M. (1988). Capturing the Capture Lab concepts: A case study in the design of computer supported meeting environments. *CSCW '88 Conference Proceedings* (pp. 257–270). New York: ACM Press.

Mantei, M., Baecker, R., Sellen, A., Buxton, W., Milligan, T., & Wellman, B. (1991). Experiences in the use of a media space. *CHI '91 Conference Proceedings* (pp. 203–208). New York: ACM Press.

McGuffin, L., & Olson, G. (1992). *ShrEdit: A shared electronic workspace* (Tech. Rep. No. 45). Ann Arbor, MI: Cognitive Science and Machine Intelligence Laboratory.

Mills, H. D. (1968). *Chief programmer teams: Principles and procedures* (Rep. FSC 71–5108). Gaithersburg, MD: IBM.

Mudge, J. C., & Bergmann, N. W. (1993). *Integrating video and large-area displays for remote collaboration* (Tech. Rep. No. TR93–2). Flinders, Australia: Flinders University School of Information Science and Technology.

Mullender, S. J. (Ed.). (1989). *Distributed systems.* New York: ACM Press.

Mullender, S. J. (Ed.). (1993). *Distributed systems* (2nd ed.). New York: ACM Press.

Neuwirth, C. M., Kaufer, D. S., Chandhok, R., & Morris, J. H. (1990). Issues in the design of computer support for co-authoring and commenting. *CSCW'90 Conference Proceedings* (pp. 183–195). New York, ACM Press.

Newell, A. (1990). *Unified theories of cognition.* Cambridge, MA: Harvard University Press.

Newell, A., & Simon, H. A. (1972). *Human problem solving.* Englewood Cliffs, NJ: Prentice-Hall.

Nisbett, R. E., & Wilson, T. D. (1977). Telling more than we can know: Verbal reports on mental processes. *Psychological Review, 84,* 231–259.

Nunamaker, J. F., Dennis, A. R., Valacich, J. S., Vogel, D. R., George, J. F. (1991). Electronic meeting systems to support group work. *Communications of the ACM, 34*(7) 40–61.

Olson, G. M., Olson, J. R., Carter, M. R., & Storrosten, M. (1992). Small group design meetings: An analysis of collaboration. *Human-Computer Interaction, 7*(4) 347–374.

Olson, J. R., & Nilsen, E. (1988). Analysis of the cognition involved in spreadsheet software interaction. *Human-Computer Interaction, 3*(4) 309–350.

Patterson, J. F., Hill, R. D., & Rohall, S. L. (1990). Rendezvous: An architecture for synchronous multi-user applications. *CSCW '90 Proceedings* (pp. 317–328). New York: ACM Press.

Peck, V. A, & John, B. E. (1992). Browser-Soar: A computational model of a highly interactive task. *CHI '92 Conference Proceedings* (pp. 165–172). New York: ACM Press.

Root, R. W. (1988). Design of a multi-media vehicle for social browsing. *CSCW '88 Conference Proceedings* (pp. 25–38). New York: ACM Press.

Rumelhart, D. E., & McClelland, J. L. (1986). *Parallel distributed processing: Explorations in the microstructure of cognition.* Cambridge, MA: MIT Press.

Schnase, J. L., Leggett, J. J., & Hicks, D. L. (1991). *HB1: Initial design and implementation of a hyperbase management system* (Tech. Rep. No. TAMU 91–003). College Station, TX: Texas A & M University Department of Computer Science.

Sellen, A. J. (1992). Speech Patterns in Video-Mediated Conversations. *CHI '92 Conference Proceedings* (pp. 49–59). New York: ACM Press.

Shackelford, D. E., Smith, J. B., & Smith, F. D. (1993). The architecture and implementation of a distributed hypermedia storage system. *Hypertext '93 Conference Proceedings* (pp. 1–13). New York: ACM Press.

SMART Technologies, Inc. (1993). *SMART 2000 conferencing system. User's manual.* Calgary, ALB, Canada: SMART Technologies, Inc.

Smith, J. B., & Lansman, M. (1989). A cognitive basis for a computer writing environment. In B. K. Britton & S. M. Glynn (Eds.), *Computer writing aids: theory, research, and practice* (pp. 17–56). Hillsdale, NJ: Lawrence Erlbaum Associates.

Smith, J. B., & Lansman, M. (1991). *Cognitive modes and strategies for writing* (Tech. Rep. No. TR91–047). Chapel Hill, NC: UNC Department of Computer Science.

Smith, J. B., & Lansman, M. (1992). Designing theory-based systems: A case study. *CHI '92 Conference Proceedings* (pp. 479–488). New York: ACM Press.

Smith, J. B., Rooks, M. C., & Ferguson, G. J., (1989). *A cognitive grammar for writing: Version 1.0.* (Tech. Rep. No. TR89–011). Chapel Hill, NC: UNC Department of Computer Science.

Smith, J. B., & Smith, C. F. (1987). *A strategic method for writing* (Tech. Rep. No. TR87–024). Chapel Hill, NC: UNC Department of Computer Science.

Smith, J. B., Smith, D. K., & Kupstas, E. (1993). Automated protocol analysis. *Human-Computer Interaction, 8*(2) 101–145.

Smith, J. B., & Smith, F. D. (1991). ABC: A hypermedia system for artifact-based collaboration. *Proceedings of Hypertext '91* (pp. 179–192). New York: ACM Press.

Smith, J. B., Smith, F. D., Calingaert, P., Holland, D., Jeffay, K., & Lansman, M. (1990). *UNC collaboratory project: Overview* (Tech. Rep. No. TR90–042). Chapel Hill, NC: UNC Department of Computer Science.

Smith, J. B., Weiss, S. F., Ferguson, G. J., Bolter, J. D., Lansman, M., & Beard, D. V. (1987). WE: A writing environment for professionals. *Proceedings of the National Computer Conference '87* (pp. 725–736). Montvale, NJ: AFIPS Press.

Spector, A. Z., & Kazar, M. L. (1991). Uniting file systems. *Unix Review, 7*(3) 61–70.

Stefik, M., Foster, G., Bobrow, D. G., Kahn, K, Lanning, S., & Suchman, L. (1987). Beyond the chalkboard: Computer support for collaboration and problem-solving. *Communications of the ACM, 30*(1) 32–47.

Streitz, N., Haake, J. M., Hannemann, J., Lemke, A., Schuler, W., Schutt, H., & Thuring, M. (1991). *SEPIA: A cooperative hypermedia authoring environment.* Darmstadt, Germany: GMD–IPSI Technical Report.

Vygotsky, L. S. (1962). *Thought and language.* Cambridge, MA: MIT Press.

Vygotsky, L. S. (1978). *Mind in society: The development of higher psychological processes.* Cambridge, MA: Harvard University Press.

Vygotsky, L. S. (1987). *The collected works of L. S. Vygotsky.* New York: Plenum.

Walsh, E. (1985). Overview of the Sun network file system. *USENIX (Winter) Conference Proceedings* (pp. 117–124). Berkeley, CA: USENIX Association.

Watabe, K., Sakata, S., Maeno, K., Fukuoka, H., Maebara, K. (1990). A distributed multiparty desktop conferencing system and its architecture. *Proceedings of the Ninth IEEE Annual International Phoenix Conference on Computers and Communications* (pp. 386–393). Washington, DC: IEEE Press.

Woods, W. A. (1970). Transition network grammars for natural language analysis. *Communications of the ACM, 13*(10) 591–602.

Yourdon, E. (1989). *Structured walkthroughs* (4th ed.) Englewood Cliffs, NJ: Prentice Hall.

Author Index

Subject Index